THE CUSTOM COWBOYS OF HOUSTON COUNTY

A Cowboy's Stories of Wild Cow Tales

BY CD LANGHAM

To Chuck & Aneita
It's been A heck of
A ride
CD Langham

Cover Design, Book Design & Editing by:
Langham Publishing
Crockett, TX

I have tried to recreate events, locales and conversations from my memories of them. In order to maintain their anonymity in some instances I may have changed or left out the names of individuals and places, I may have changed some identifying characteristics and details such as physical properties, occupations and places of residence.

This book is based on actual stories by the author and friends. Cowboy work is dangerous. Please don't try this at home.

Graphics source: openclipart.org/unlimited-commercial-use-clipart

DEDICATION:

To my wife Pat, with love,
who without her insistence and inspiration,
this book would have never been written.

A SPECIAL THANKS

To my daughter-in-law Natalin Langham,
for all her help in putting this book together.
Without her help, I could have never got it done.
"Thank you Natalin"

~ CD

PREFACE

This book is about the Custom Cowboys of Houston County. It's a hard life. We've lived our lives fighting the rain, wind & weather, while riding the brush, briers and thickets of east Texas.

The men these stories are about made a living with their horses and dogs during the Jet Age. They play the most important role in the cattle industry. No way ranchers could do without them. Their jobs are penning cattle to work, de-worming, helping with their necessary vaccinations, catching wild cattle, pasture doctoring sick cattle, pulling calves, thrashing cattle out of the brush and many other things that people never think about. They also get cattle out of the backwater when rivers & creeks overflow, trailing them down highways and roads when there is no way to haul them.

I, for one, am glad I got to serve the public with my services as a cowboy for all the years I did. These are my stories, plus some that have been told to me by other cowboys. They started out as true stories, but the truth may have been altered to some extent.

HOUSTON COUNTY CUSTOM COWBOYS

- Todd Merriweather
- Butch Arnold
- James Ashley
- Henry Beaver
- Billy Bob Christian
- Billy Crowson
- Ronnie Goolsby
- Mark Avery
- Rick Hargrove
- Jonny F. Harrison
- C.D. Langham
- Jimmy McLean
- Caleb Rhone
- Brandon Richburg
- Johnny Shroyer

There are many more not mentioned here,
as you will see in the pages that follow.

TIME

Time is the element we all must obey
Death is the price we all have to pay
"God" holds the keys to how long that we last
But we hold the keys to what we did with the past

So, we must move along, we must make haste
For we surely don't have any time to waste
This time that we have passes so fast
So we must live for the future and not in the past

Life is so short, how the days fly by
How tall are the mountains, how big is the sky
These questions I ask as I'm passing through
What went with the time, how fast that it flew

Life's problems and troubles can sure wear you down
But I know where the answers can always be found
All of these truths that you've ever heard
Can be found in the pages of "God's" Holy Word

When sin comes against me I'm not very strong
But may "God" keep me from all wrong
My soul that was lost, my Jesus did save
So why dwell on the sins that "God" has forgave

- CD

THE OLD COWBELL

My family and myself ran cattle in timber company land that joined our property for many years. In 1963, I had just came back from Idaho, where I had worked on a ranch that summer.

We had cattle in the woods, and I discovered that one of the bell cows had lost the clapper out of her bell. She was a big high horned cow, about ½ Brahman. One day I put a big square nut and a piece of wire in my pocket to use for a clapper.

I just happened to catch her out in the opening where I could get a run at her. Well, I busted her out away from the woods and roped her. I stepped off my horse and it jerked her down, but before I could get on her head she got up. She ran at me, and the rope got half hitched around my boot and spur. My foot was pulled up between her front legs and I was dragging under her.

Finally my boot came off and I got away. I was pretty close to a big Sassafras tree, so I got back on my horse and drug her over to the tree and tied her. Then I took the bell off and fixed the clapper. The bell collar was leather and a few years later it broke and she lost the bell after all.

Forty years went by and a logger found the bell and left it with my neighbor. He stopped me and asked if I had lost a cowbell. I thought about it a minute, then asked if it had a square nut for a clapper. To my surprise, he said it did.

Today that bell hangs in front of the house and rings when anyone opens the gate. It's a part of history no one knows the story behind but me and you!

- CD

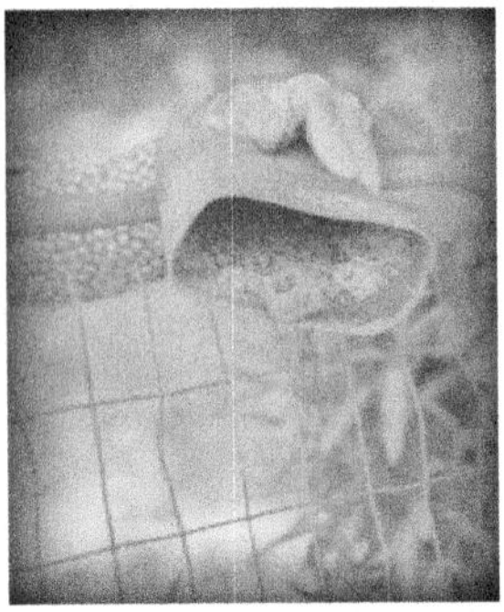

CD Langham & The Old Cowbell

HENRY BEAVER AND HIS DOG

Henry had a big Airedale dog that he couldn't keep from going to cattle before he sent him. He kept him on a chain when we'd go work, because he would keep cattle bayed somewhere if he was loose.

One day he was going to pen cattle, so he had a plan. He was going to keep him on the chain until he got to the cattle. Well, he had his horse saddled, and he took the dog on the chain out to the horse. His rope was coiled around the saddle horn so he tied the chain to the end of the rope. He stepped on his horse, the dog went under the horse, and around the horse's legs.

Needless to say, the horse broke into and threw Henry flat on his back. The horse took off and drew the rope down on the saddle horn, then ran away with the dog on the end of the chain. Just about the time Henry sat up, the dog hit him square between the eyes, rendering Henry and the dog both unconscious.

Never tie your dog to your horse. **~Henry Beaver**

A JD JONES BULL CATCHING STORY

J.D. Jones had an Angus bull in Coldspring Hunting Club, which had 34000 acres in it. Rick Hargrove and myself went to see if we could find him. J.D. knew where he was last seen, so that's where we started looking.

We rode a long time before we found any tracks, but when we found a track the dogs acted as if they could smell it, so we sent the dogs and the race was on. Rick heard the dogs coming, so he backed up in the in the brush and waited.

The bull was coming down an old log road, and when he passed, Rick fell in behind the bull and roped him. We were miles from the truck by this point. So, we got two ropes on him, one of us got behind and one in front. We finally got him back to the truck.

This bull had been gone for a year. They was sure glad to get him back.

~ CD

A JOHNNY SHROYER SNAKE STORY

Johnny worked for Granada Ranch for a number of years. He had many experiences while working there. One that stood out to me, was dealing with embryoing cattle. They sometimes embryoed as many as 500 a week.

Johnny had killed a chicken snake in the feed room that morning. He was bringing cattle down the chute for the vet to palpate. He thought how funny it would be to put that snake in one of those cows that the vet was going to check.

So, he lubed up the snake good and put him in the cows rectum. Well, the vet ran his arm in the cow and found there was something in there that shouldn't be.

He began pulling the snake out, and it was folded up pretty good. Directly it came out and fell on the floor. The vet was so startled he turned to run, but behind him was a heavy Powder River panel and he ran into it so hard it knocked him down flat. He left there crawling on his hands and knees quick as he could, scrambling to get away from that snake!

~Johnny Shroyer- J-S

Johnny Shroyer

A RONNIE GOOLSBY STORY

Ronnie Goolsby and Carroll Langham were penning some cattle for John Powell. Ronnie was riding a young horse that he had just bought from James Horton. They were riding off from the lot to pen the cattle. Ronnie said, “I just bought this horse and I sure like him. James has roped a few steers on him in the roping pen.”

Carroll had a yellow black mouth puppy that wasn't broke good yet. He had a loop ready in case the puppy tried to go to cattle before he was ready. Sure enough the puppy went by him in a hard run. Carroll roped the puppy when he went by. The puppy ran towards Ronnie's horse. Carroll took off to the right to try to keep him from going under Ronnie's horse. Well, the dog came to the end of the rope and it slung him into Ronnie's horse's back legs.

The horse broke into a buck. Ronnie made a pretty good bronc ride, but he finally bucked off. The horse ran away, stepped on the reins and broke his head stall.

Carroll said, “Are you alright?” Ronnie replied, “I am just tired, just let me lay here a while.” Carroll said, “I don't know if that is a rope horse or not.” Ronnie said, “I don't know about a rope horse, but he sure is a bucking horse.”

~Ronnie Goolsby G

INCIDENT AT GENE HARRISON'S

Carroll and me, Rick Hargrove, went to catch a cow that had got out of Mr. Harrison's pasture. We finally found the cow, and drove her back to the pasture, but she decided to make a run for it just before we got her back in.

Carroll roped her and we drove her back to the trailer. I was going to get another rope on her and did, but she got tangled up, so I threw my rope down to get her untangled. When the ropes got straight, I got off my horse to pick it up. She made a run or two at Carroll and finally stopped.

When I picked up my rope to get back on my horse, she made a run at me. I was in a bad position and she had a lot of slack between her and Carroll. I was outrunning her until she hit the end of the rope and swapped ends. Her back end hit me like an 18 wheeler, knocking me about 30 ft.

Carroll nearly fell off his horse laughing, but I didn't think it was real funny. Anyway, it didn't turn out too bad...we caught the cow, and I didn't have any broken bones. **~Rick Hargrove H-H**

Rick Hargrove

A BULL CATCHING STORY

Dick Murchison had a Brahman Limousin bull in Denny Graham's pasture. Myself and Billy Bob Christian went to catch him. Billy Bob was riding a *Doc's Quixote* horse that was pretty broncy. Amber, Billy Bob's 10 yr old daughter, had been riding a good buckskin horse of mine, and it was in his trailer, but he wasn't saddled.

The bull was near some big Pine trees. We ran him through the strip of Pine trees and when he came into the opening on the other side, I roped him. He was a pretty big bull, and I was having trouble stopping him.

I noticed my horse's neck was getting shorter and suddenly my saddle came off over his head. I had just bought that longhorn saddle and put saddle string on it. That bull left with my saddle bouncing ten feet in the air. Billy Bob couldn't get his horse even close to the bull with that saddle dragging. He unsaddled his horse and put his saddle on the buckskin horse that was on the trailer.

The bull was way down Elkhart Creek by the time I caught up with him. There was a big willow tree with a limb hanging down to the ground. The saddle had went up the limb and lodged in a fork ten foot off the ground. The horn knot had slipped off the saddle horn and the bull was headed back to the highway before I caught him again.

My saddle was torn up pretty bad and the strings I had put on it was tore off, plus my extra rope tied on it was gone. Anyway, we got the bull and carried him home.

- CD

WILD HORSES

Doc Ashley bought two horses from Billy Gibson, range delivery (that means, "you bought 'em, you catch 'em"). There were two horses, a four and a five year old stallion. One was black and the other one bay. Doc called and wanted me to catch them.

I got through where I was working about 3 o'clock one hot summer day. I went over there to catch them. On the way over there I had a plan. I've never rode in a pasture with a stud horse that didn't run up and check me out. I had me a loop made and was ready.

When the horses ran up, the black was in front and I chunked my rope on him. He had never had a rope or a human hand on him. He turned over backwards probably five or six times before he finally choked down. I threaded my rope through the trailer and pulled him in.

Then I set out to catch the bay. I tried to get close enough to rope him, but he stayed a long way away from me. There was two separate pastures with a cross fence and a gap between them. There was an old truck tire laying on the ground, so I took my long rope and tied it to that truck tire.

I made a big loop and hung it across the gap with the bottom of the loop about 2ft from the ground. Then I got behind the horse and run him toward the gap. When he got to the rope he took his nose and raised the rope up and went under it.

Then I got me another long rope, made me a loop and got some paper out of my truck and tied that loop to the bottom of the other loop. Then I got after him again, and run him toward the gap. This time he stuck his neck in the bottom loop. The pasture on the other side of the gap had Post Oak trees scattered all down the side of the hill. That tire was dragging behind the horse, and he was plenty scared. The truck tire would go around one of those trees and come out from around the next one, and then go around another. It would nearly jerk him down.

He ran all around that hill and when he came to the cross fence, he jumped it. When the tire got there it uprooted about 6 or 7 posts. It pulled the five wires a hundred foot out in the pasture. I ran up and threw another rope on him. I left my horse and the tire holding the horse and I went and got the truck and trailer and loaded him.

When I got to Doc's he wanted me to put a halter on each one of them and that wasn't easy. I backed to a barn that had a solid wall up and unloaded them in the barn. He made a deal with two black men to break one for one, but they didn't get either one of them broke. They both ended up at Calhoun's kill plant. ~ C_D

A COW CATCHING STORY

James Ashley and myself went to catch a cow. The cow was in some brush when we found her. James went in the brush and ran her out. He built to her to rope her, and I went ahead and made a loop in case he missed.

The cow stumbled about the time he threw his rope, and the rope just missed her. I was right behind him ready to rope and his rope was laying on top of the grass.

My horse picked James' rope up on his front feet, when the rope came tight, it turned my horse a flip. James said the only thing you could see sticking out from under the horse was my hands and feet.

I thought I'd never breathe again, but when I finally got some air back in me, I'm glad to say I was alright. We caught the cow, loaded her, and carried her home. Thank you "Lord". ~ C_D

Horned Cow

ANOTHER BULL CATCHING STORY

West of Pennington there is a little community once called Four Dice, now named and pronounced Fodice. An old black man by the name of Napolean Vonner had a place that was 400+ acres and a bull that he couldn't catch.

Rick Hargrove and myself went to see if we could catch him. The place had been logged and grown back up in bushes and brier vines. A black cowboy had been trying to catch this bull for three years, without success.

The first time we went we bayed the cattle up in that thicket, and he slipped away from the cows, the dogs would keep baying cows but we couldn't find the bull. The next time we went, was about 4 o'clock pm. New Year's Eve. We stopped and talked to Napolean and he said, "You think you can catch him before dark? L.D. Patton couldn't catch him in three years." We said we would like to see if we can. Napolean said, "My wife just fed the cattle in that little opening by the road."

Rick and I decided to unload the horses and dogs and ride down the road as close as we could, then send the dogs and hope the bull would stay with the cows until we could get there. The cows saw us before we got to them, and left. We sent the dogs and they stopped the cows in the edge of the woods. Rick was on one side of the cattle and I was on the other. Rick said, "He is coming out on your side."

I saw the brush parting and I took to the bull. I was riding a good brush horse and I was able to keep the bull in sight until I got to White Rock creek. There was a pretty good opening across the creek and I was able to rope him in that little opening. He was a high horned, ¾ Brahman bull.

There was a long tall Pine tree in the middle of the opening when my rope come tight, the bull came to my horse and I had to outrun him around and around the tree. Finally my rope was halfway around the tree and the bull could nearly reach my horse with his horns.
I could still hear Rick trying to get the dogs off the cows half a mile away. I hollered and hollered and finally he heard me. When he got to me, he got another rope on the bull and drug him to a big red oak tree. We tied him off to some big limbs on the red oak and started hunting a way to get to him with a truck and trailer.

We finally got to him with the truck and trailer and drug him in the trailer with the horses. The bull was in the front of the trailer and the horses in the back with a gate separating them. When we got out to the highway, I could tell there was something going on back in the trailer. The bull had bowed the gate to where the gate was against the horses. It was still a long way to East Texas Livestock Co. sale barn.

About halfway there, the highway patrol stopped us. He said, "The reason I stopped you is because you don't have lights on your trailer." Rick said, "I had some until I drug the wires off on all them bushes, but I will get them burning." The highway patrol said, "If you can, I'll let you go."

The bull was about to turn the trailer over trying to get to the horses, and he had broke the hinges off the trailer gate. We tied the gate back to the bow with a lariat rope. The highway patrol said, "Just go on and get out of here!"

We got to the sale barn, and put him in a pen in the middle of the barn so he would have to jump several fences to get out. It was a week before the next sale, and the bull was so wild he didn't eat or drink and he walked a trail a foot deep around the pen before the next sale. When they ran him through the ring, I bet he had lost 300lbs. ~ CD

Napolean Vonner Bull

JOHN POWELL'S SANTA GERTRUDIS BULL

Walter Sherrouse and myself were penning cattle for John Powell. We had the cattle bayed up on the side of a hill, when a Santa Gertrudis bull broke out of the cows and headed for a 40 acre thicket.

I ran up behind him and roped him just before he got to the thicket. I looked up and five cows had broke and were headed for the same thicket a little farther down.

We needed the cows worse than we needed the bull, so I threw the rope down an headed the cows before they got to the thicket. We got the cattle penned and worked, then we went back to get the bull that had my rope on him. He was in that 40 acre thicket.

We spent the evening trying to get him out, but never could get him out. We had left the cows on the other side of the cross fence which was a good ways from the thicket. I told Walter, "We will come back early in the morning and he will be over by the cows."

The next morning we left he truck and trailer up on the road so we didn't spook him and rode to the top of the hill very quietly. Sure enough, he was over there by the cows. We slipped as close as we could, he saw us, and made a run for the thicket. I roped him just as he got to the woods. When Walter saw that I had roped him he took off to get the truck and trailer.

By the time I got the bull stopped he was out in the woods a little ways. I thought I would pull him back out in the opening before he got tangled around the trees. The horse I was riding was a young horse, and not very big. I thought, this little horse can really pull, but then I noticed the bull was under my horse. About that time, he threw my horse end over end. I landed right in front of the bull on the ground.

The bull scooped me up with his horns, and threw me way up in the air. When I came down, I landed on his neck. Needless to say, you can't ride a bull's neck very far.

When I landed again he started rolling me back to the woods. He rolled me up in some briers and quit me. I just layed still for awhile, and he stood there and looked at me. I didn't think he'd let me stay there forever, so I crawled on my hands and knees behind the briers.

About that time, Walter got back with the truck and trailer. He had been riding an own son of Cutter Bill, but he was pretty lazy. I got Walter's horse, and threw another rope on the bull. I was going to pull the bull over to the trailer, but the horse didn't want to pull. The bull ran and hit the horse in the back end. The horse ran, but the bull stopped.

When the horse came to the end of the rope it jerked Walter's saddle horn out by the roots. It hit me in the belly and knocked the breath out of me. We took my horse and loaded the bull, but Walter was pretty upset that I had broke his saddle. He had a new tree put in his saddle, and he's still riding it. - CD

Santa Gertrudis

A BYLER BULL CATCHING STORY

Larry Brasher and myself went to find a bull for L.P. Byler. He bought him from Clark Angus Ranch in Huntsville and put him on a place near Augusta. We tracked him through two places that had cattle on them, but you could track him easy because his track was twice as big as any of the other cattle's track.

We found where he went out on the county road, then off in Pedro Creek and swam out on the other side. That put him in Tommy Cutler's pasture which is 900 acres of mostly woods. We rode most of the day before we found him. He was in a horseshoe bend on Pedro Creek. I told Larry if he would run him out of the brush, I would rope him.

There was a light line right of way that was probably 200ft wide. When he came out in the opening, I ran up behind him and roped him. By the time I got him stopped he was nearly to a cross fence. Larry got a heel rope on him and tied off to the bottom of a big corner post.

I rode back about 3 miles to where we left the truck and trailer. I went up to Tommy's house and he told me the best way to get back to where the bull was tied. So, I drove around and backed the trailer up close to the bull, I threaded my rope through the trailer to pull the bull in the trailer. I told Larry to untie his heel rope from the corner post. When I got the bull nearly to the trailer, he turned and took off the other way.

It pulled my horse up on the side of the trailer and he just hung there for a little while. Then he sort of flopped around a little and the saddle came off and wedged between the bars on the trailer. Larry tied his rope back off and we backed up enough that we could get my saddle out. I re-saddled my horse and we put another neck rope on him and both horses was able to load the bull. Mr. Byler carried the bull back to Mr. Clark and traded him for a gentler bull. ~ CD

L.P. Byler's Bull

A PREACHER MCQUEEN BULL CATCHING STORY

Buck Jones saw after Preacher McQueen's cattle for him. Buck called me and said the Preachers bull was out in the paper company woods around the Houston County Lake. There was 18,000 acres of woods around the lake. The bull was with a little bunch of cows that belonged to Bill Ike Lowe. Buck wanted to ride over there with me when I went to catch him.

Saturday morning early, I went by and picked Buck up. He had been drunk the night before and he didn't look too good. We parked the truck and trailer near where the cattle had been ranging. I hadn't rode very far when I saw fresh tracks. I sent the dogs and they went to baying not far from where I sent them. When I got there, the bull was with Bill Ike's cows. I drove the whole herd through a little old thicket and out in a big opening near the old Deitz house.

Then I noticed the bull wasn't with them. I called the dogs and went back in that thicket and the bull had hid in some brush. When he saw that the dogs had found him, he tried to make it back to the cows. The cows had moved on out in the opening but I roped the bull before he got back to the cows.

I tied my reins around the rope and walked back where the truck was. Buck was sitting in the truck asleep. I tapped on the window and woke him up. He said, "What did you do fall off your horse?" I said "No I left him tied to the bull." Buck said "That bull will kill your horse". Buck showed me how to get my truck and trailer over to where the bull was. When we got to where the horse and bull was, they were both gone. I guess the bull took after the horse and ran him way over the hill nearly to another set of woods.

When we got to where we could see them, one of them was on one side of a Pine tree, and the other on the other side. I told Buck to let me get on the trailer with another rope and for him to get me close enough to get another rope on him. He did, and I already had the end of the rope tied to the trailer. Once I roped him, I went and got on my horse threaded my rope through the trailer and pulled the bull in the trailer. Buck said I would never believed that little horse could have loaded that big bull. What Buck didn't know was, that the bull led up on the trailer real easy.

- C_D

A DENNY GRAHAM BULL CATCHING STORY

My wife and I hadn't been together very long and she didn't know a lot about catching bulls. I got a call to go catch a grey Brahman bull that had killed Denny Graham's bull. He had crossed the highway, went through several places, and ended up in an old woods pasture near Doyle Bennett's.

He was with a little bunch of cows in that pasture. We unloaded our horses and rode through some woods and across a tank dam. When we saw the cattle standing on a hill in the opening, I sent the dogs and they bayed them on the side of that hill there. I told my wife, "When I get him roped off this line backed horse, you be there on that stud horse and jump off and let me have him. "

Things worked as planned and I had him caught between the horses so they couldn't get hooked. I told my wife to sit on the stud horse and don't let the horses get on the same side or the bull could hook them. I walked back and got the truck and trailer, and when I came back she was out there behind a tree. She came and got the truck, I ran and jumped on the stud horse. The bull took off, when he hit the end of the rope it hit the saddle sideways, and jerked a whole chunk of the tree out of the saddle right where the riggin was screwed into the tree.

The saddle went one way and I went the other. My wife had backed up the trailer by then. I took my knife out and cut the rope right at the saddle horn, ran in the trailer and tied that rope off. Then I took the other horse, and loaded the bull. I closed him off in the front twelve foot of the trailer, and loaded the horses in the back eight foot.

The bull went down the side of the trailer hanging his horns under the bars of the trailer and hooking up on it. It put so much pressure on the floor, it broke all the boards out of the front of the trailer and he was walking around on the ground. I unloaded the horses and tied his head up in the back of the trailer.

I managed to get to the sale barn and unload him before he tore anything else up. I broke one $600 saddle, the $150 of floor out of the trailer, one $30 rope and gave my wife the diarrhea.

I got $75.00 for catching the bull.

~ CD

A WILD COW CATCHING STORY

Rick Hargrove and myself went down to George West Texas to catch some wild cattle. This man had bought 3300 acres of that Mesquite country to make a game ranch. He had cleaned a right of way around it for his high fence and cleaned up two little patches for food plots and the rest was solid Mesquite. These wild cattle was eating his food plots faster than they could grow. They flew over it twice in a helicopter and counted twenty head of grown stuff and few calves.

The sale at Three Rivers was on Wednesday, so we went out on Monday night. We spent the night in a tin bunk house and nearly froze to death. The next morning we were saddled up by daylight, and the man's son led us over to the first food plot. There was ten head of cattle on the food plot, so we sent the dogs not knowing what they were going to do, because they had never saw dogs before. Surprising to us they bayed up pretty good. We each had five ropes tied on our saddles and Rick's Uncle Van was horseback just to carry more ropes.

We caught eight out of ten before they got out of the food plot. Two cows got in the brush, and I had a good find dog but it was so dry she couldn't trail them.

We split up so we could cover more ground, and I found seven head in a bunch with my dogs. I only had five ropes but I caught five head and tied with one rope. The last one I roped was a big white bull yearling that weighed about 700lbs. He was a muley and I couldn't tie him off or he would choke to death. I sat there and held him for two hours before Rick and Van showed up. There was still one that wasn't caught, but she had stayed with the ones I had tied, so Rick roped her and we had all seven of them.

We found one more red bull two year old and caught him. By then it was dark and we had seventeen tied all over that 3300 acres. We were way up in the night hauling them to Three Rivers to the sale barn.

There was 3 left, but I didn't hear from the man again so I guess he did something with the other three. Seventeen out of twenty wasn't too bad, although, I would like to have gotten them all. - CD

CD & THE BLACK MAN'S BULL

One Tuesday morning, I pulled up to the sale barn in Crockett. I had a load of cows on my trailer, and did not have room for my horse. Just after I got stopped in line, Carroll walked up to my truck and asked me where my horse was. I told him where I had left him. He said he needed to go catch a bull and wanted me to go help him. I told Carroll that as soon as I got unloaded I would go get my horse.

We sat there a few minutes and Carroll said, "I'll tell you what, knowing Cecil, the bull won't be very big, so when you get unloaded just go with me and bring the truck when I get him caught."

I told him that would be fine. When I got unloaded I pulled around and parked my truck. I got in the truck with Carroll and we followed the black man to the pasture where the bull was. We pulled inside the gate, and Carroll unloaded his horse. Me and the black man sat in the truck while Carroll rode in to the woods where there the bull was supposed to be.

In a few minutes, here comes a huge Beef Master bull out of the woods. This bull weighed at least 2000 lbs. When Carroll came out of the woods, he took to the bull and since he was riding Old Buck, he was not catching up too fast. The bull went to making a circle going back to the woods.

When Carroll got close enough to rope, the bull was just about to stick his head back in the woods. Needless to say, he caught the bull. When Buck saw what was ahead of him, he sat his butt on the ground trying to stop the bull. When the bull hit the end of the rope Old Buck, left out like a rocket.

When Old Buck left the ground, Carroll was just like he was sitting in mid air for a split second. Then he left behind Old Buck. His left foot was hung in the stirrup. Just as Old Buck hit the tree, Carroll passed him up and it popped him like a whip.

Old Buck was in the air when he hit the tree, which stopped him and the bull. The horse fell to the ground and the bull was spun around another tree and jerked down. The black man was sitting in the truck with me and he started yelling "He's dead! He's dead!" And I said, "No he's not dead, I see him kicking".

I cranked the truck and headed to where they were. As I got to them, the bull was getting up, and trying to get to Carroll. Carroll was kicking loose from the horse and trying to get out of the bull's reach. The bull couldn't get to Carroll because Old Buck was still hung on the tree. I got out of the truck and got some more ropes out. I got another rope on the bull and got him tied to a tree.

When Old Buck got up, he barely could walk from hitting the tree. We got some more ropes tied together and ran them through the trailer and tied off to a tree. When we got the bull lined up with the trailer I drove forward and pulled the bull in the trailer. Old Buck was never the same.

The moral of this story is "Don't never judge the bull by the man that owns him." **~Billy Bob Christian**

Billy Bob Christian

Beefmaster Bull

A LOST CATTLE STORY

Floyd Hicks had 10 cows and 10 calves get out in 18000 acres of paper company land around the Houston County Lake. They were gone all summer and part of the winter. I had looked for them several times, but was not able to find a single track, not even a cow pile. I had people watching for them that lived in the area, but no one had seen them. The man that owned the cattle called Christmas Eve and said they were seen bedded down on a dirt road near the lake.

Christmas morning I went to where they were seen. I got there at daylight hoping they may be bedded down in the same place. They weren't, but the dog could still smell them. I sent the dogs and they went over a couple of hills and started baying them. It was a five or six mile drive back to the pasture through the paper company woods. It took 5-7 hours to drive them back to the pasture.

On the way, I discovered there was a fresh brand on everyone of them and there was no calves on them. Come to find out, the cowboy that worked the cattle on the Spence place, had put the brand on them, and the man that *thought* he owned the cattle, sold all the calves off them. Then he drove them back out in the paper company land. Floyd Hicks filed charges, but never got anything for his cattle.

Turned out to be a rustler story. - CD

Rustled Cattle

A MAD COW STORY

I day-worked on the 7A Attabury Ranch, doctoring e-coli scour calves. Bob Thompson was the ranch manager, and a good friend that liked to help everybody.

I drove up one morning to doctor scour calves and Bob had a horse saddled for a man that had just moved here from Michigan. His name was Gary Miller. Bob said that Gary needed to work and he was going to let him help me doctor scour calves. He didn't know much about cattle, but he could ride pretty good.

When I got a calf roped, his job was to get the calf down while I drew the medicine to doctor the calf with. It worked fine until about the 5th calf. He got the calf down and here came the cow. He turned the calf loose and ran for his horse. I told Gary if he would stay on the opposite side of the calf from the cow, she wouldn't bother him. He asked me if I was sure. I told him for sure that she wouldn't.

He got the calf down again, and here came the cow again. She never slowed down when she got there. She hit him and rolled him 20 ft. He jumped up and ran for his horse, but about the time he got hold of the saddle horn, the horse left. Someway he mounted his horse doing about 30 mph.

When we got back to the barn, at dinner time, Bob Thompson loaded us in that old white Ford Bronco to carry us to dinner. Bob asked Gary, "Well, did old Carroll teach you anything today?" Gary responded, "He taught me one thing, he sure will lie to you." ~ CD

CD Langham at Attabury

ESCAPED CATTLE STORY

Mark Avery and myself went near Alto to get back some escaped cattle. There was 10 Simbra heifers and 1 calf that had been missing for six months. The man that owned them had found cattle sign in 34000 acres. The Coldspring Hunting Club. We all had twenty four foot trailers, so we found a wide place in the log road and pulled 2 trailers off to the side. We backed the other up to those two.

We had a funnel made with the other two trailers. We set out to find the cows. We rode until about 11 o'clock before we found any cow sign. We finally found where they had bedded down the night before. The dogs could faintly smell the cattle, so I sent the dogs and they left in a run.

We rode hard trying to keep up with the dogs. They finally bayed the cattle a mile and half away. We drove the cattle several miles back to where we had the trailers set up. When we started to drive the cattle into the funnel we had made with the trailers, two of the cows broke out and left. One of them stayed in the log road long enough for me to rope her. I tied her off to a tree and went back to help Mark.

We drove the eight cows and one calf up in the trailer. Then we set out to find the other cow. There was a dog left with her, so we thought we may hear him barking, but we rode the rest of the evening and couldn't find her. The man that owned the cattle and his friend rode the log roads stopping once and a while listening for the dogs. They went out to the Wells highway which was about 10 miles away and turned back toward Alto. They stopped to listen for the dog and they heard a dog in a barn, not far from the highway.

The man said, "I hear them dogs in that barn up yonder." The friend said, "That is someone else's dog." But the man that owned the cattle said, "I've been listening to them dogs all day, and I would know them anywhere." They got out of the truck and walked up to the barn, and sure enough, the dog had the cow bayed in the barn. It had a little small door on it that was open. They closed the door, went and got the truck and trailer, backed up to the barn door and ran her up on the trailer.

It was dark when they got back where me and Mark were, and we didn't see the cow on the trailer. We told them we had rode all evening and hadn't found her. They said, "There she is on the trailer." The cow was almost back to Alto, ten miles away. The dogs had her bayed in the barn all day.

~ CD

A BILLY CHRISTIAN BAD HAT STORY

Billy Bob Christian and myself had hauled some cattle to the sale barn and ate dinner while we were there. When we came out, there was a man selling black western hats. He had them layed out on the back of his truck and he asked if I wanted to buy one. I picked one up and looked at it, and it had Resistol wrote in the hat band. I said, "How much?" He said, "$25.00." So, I bought one.

Billy Bob came out a little after me, and it had started to rain. The man was trying to put the hats up to get them out of the rain and he asked Billy Bob if he wanted to buy one. Billy Bob asked how much. The man said $25.00. Billy said, "I'll give $20." The man sad, "Okay". So Billy bought one for $20.00.

Well, it continued to rain, but we worked anyway. We were riding across the pasture to the cattle we were going to pen, and the rain was running off the hat and down our face. Billy Bob said, "You are turning black!" I said, "You are too!!" The only thing we could figure out, was the man must have bought cheap made hats out of Mexico, dyed them, and sewed the Resistol hat bands in them. All that black dye ran down our face. Good thing that man was gone when we came back. - CD

A NEW HAT STORY

James Ashley and myself were going to catch a Brangus bull for Glen Barrett. On they way, we stopped at Joe Robinson's Feed Store to get diesel and I bought me a new straw hat. We got out to where the bull was with Paul Oliver's cows. We sent the dog to bay the cows up, hoping the bull would stay with the cows.

There was a great big patch of woods on the place, and the bull left the cows and went into the woods. The dogs went with him and he finally came back to the cows. But, he didn't stay long and left again. I built to him down a long hill. Just as I roped him, he made a right hand turn. When the rope came tight it jerked that horse 20ft. The first thing that hit the ground was my new straw hat, and my head was still in it. That hat was rolled up like a snowball. There went my $35.00 hat for a $50.00 bull catching.

That's why cowboys are so rich. - CD

A JACK DONKEY STORY

Rick Hargrove and myself went to pen a set of cows for Coy Blakely. We rode horseback from the road so that we didn't spook the cattle into the woods with the trailer rattling in.

When we got to the bottom of the hill, before the cows could see us, we sent the dogs, hoping the cows would bay up in the opening. We heard the dogs baying on the cows, then suddenly we didn't hear them anymore.

When we got to where we could see them, the cows were bunched, but the dogs were nowhere around. We rode to the woods and a Jack donkey had run the dogs out into the woods.

I told Rick, "We are going to have to rope that Jack and tie him to a tree before we can pen those cows." Rick lined that donkey out across the pasture and roped at him and caught him by the ears. I knew he couldn't hold him by the ears, so I built a loop and took to him.

Just as Rick's rope came off his ears, I ran up behind him. When I roped him, the donkey quickly ducked to the right behind a huge Elm tree, jerking my saddle under my horse's belly and the donkey into the tree. It threw me thirty feet out into the pasture and knocked the donkey unconscious! We tied him to a tree when he came to, and then we penned the cows.

~CD

PREACHER JORDAN

Preacher Jordan is a cowboy and a team roper. We were discussing the gun laws they are trying to pass, and also the social security laws they've been debating on.

He said they are going to take our social security checks and our guns, then we are going to starve to death because we can't even commit suicide.

I said "Maybe you can hang yourself with one of your old worn out heel ropes." Preacher said, "As bad as I rope, I would probably miss myself."

~Preacher Jordan

ANOTHER PREACHER JORDAN STORY

Preacher stayed too long at the bar one night. On his way home, he was weaving on the road pretty good. A local officer pulled him over.

The officer said, "I thought you may be drunk by the way you were driving." Preacher said, "Man, am I"! The officer said, "Get out of the car and lets see you walk." Preacher replied, "Man, they ain't no way I can walk, if you can't tell I'm drunk, you need some officer training."

The officer asked Preacher if he could get home.
He said, "I might can." The officer said, "I better not see you back on the road tonight."

A few days went by. The next time Preacher saw the officer he asked, "Why didn't you lock me up the other night?" The officer said, "If you had told me you had two beers, I would have."

~Preacher Jordan

COWBOY'S WAGES

Rick Hargrove and myself was working on the Harrelston ranch. We were worming cows and working calves. Black legging, ear marking and cutting bull calves. We got all the small calves worked in the chute, and we were going to heel the big calves down.

James Harrelston's little boy was sitting on the fence watching us heel the calves with our ropes. He said to me, "That's what I want to be when I grow up." Then He asked, "What do you have to do to be a cowboy?" I replied, "There's nothing to it son, you only have to learn to enjoy being poor."

- C_D

A CADDENHEAD COW STORY

Bitt Caddenhead was having e-coli scour problems, and I'd went down to doctor scour calves for him several times. I went down by myself one day. The cattle was probably a mile from where I left the truck. I had rode down there where the cattle were and had my medicine in the saddle bags. I had doctored several calves already.

One calf, I hadn't been able to slip up close enough to rope. I crowded in the cattle enough to cut him out, and his mama came out of the bunch with him. She was running right by him on his right side, and I knew she would foul my loop if I threw at him while she was running along side him.

She decided she had rather fight than run, and she turned broadside to fight at my horse. I ran into her doing 30mph. We all flew in the air, my horse turning a flip, and slamming to the ground. I went off over his head and landed flat on my belly out in front of my horse.

When his back end came over, he landed right in the middle of my back. He weighed 1250lbs. (over a half ton), so he was a pretty good load. The horse and the cow were both still on the ground when I got up. Then, when the cow got up, the first thing she saw was me. She took to me, hooking me as I was running around and around the horse that was still on the ground.

Finally, the horse stood up, and I stayed on the opposite side of the horse until the cow went on. I started to get on my horse, but I couldn't pick up my left arm. So, I led my horse off in a ditch and managed to get on him. I rode back to where my truck was, and came back to Crockett.

I stopped at Joe Robinson's feed store, and called Bill Platt. He carried me to the V.A. Hospital in Temple. There, they x-rayed me and found I had a broken collarbone and two broke ribs right over my lung.

The two ribs sure hurt. They gave me a harness that laced up behind my back. My wife laced it up a couple of mornings, and helped me saddle my horse a couple of times, but I never missed a day's work. Although my collarbone never hurt, it's still separated because it never had a chance to heal.

~ CD

TRINITY RIVER OVERFLOW

In 1990, the Trinity River got bigger than it had since 1957. It was hill to hill on both sides of the river. It was probably three miles wide. We were moving cattle out of the overflow on both sides of the river. We were going to trail some cattle a couple of miles to John Merriweather's pens, but we had to cross the river bridge, which was 300-400 yards wide.

A deputy was flagging traffic. Just as the cattle started upon the river bridge, the deputy stepped out on the bridge on the other end and took off his hat, flagging cars. Those old wild river bottom cows that had hardly ever seen a man, jumped off the bridge and hit the flood waters. They were still swimming when they hit the timber ½ mile down the river.

I went home and got a motor boat. We put in the flood water, and went down to where the timber started. It was thick, because only the top branches on the trees were sticking out of the water. We floated cattle out beside the boat for three days. We had a couple pairs of nose tongs, we would float the cows out to the highway and stop just before the cows could stand. If you didn't, if they could touch the bottom, they would turn the boat over.

James Lloyd Murray would ride out and meet us on my old bay horse. He would put a rope on the cows, drag them out to the highway, and load them on the trailer.

Channel 13 news team stayed there all 3 days with their satellite van filming us floating the cattle out.

- CD

Trinity River Flood

A HORSE STORY

I bought 4 colts at the Happy Cook Horse Sale. Bob Thompson, the Attabury Ranch manager wanted one of them. I let him have one and I kept the other three.

The papers were in transit on two of the colts when I bought them, so I didn't know much about them. When I got the papers, turned out there was one of them that was bred real good. He was an own son of *Freckles San Man.*

When he got old enough to break, I broke him and he made a real fine horse. He was an extra-ordinary horse. If a cow went to fighting him, he would wheel in the air and kick her with both hind feet. He was hard to ride, but he would teach a cow a lesson.

Rick Hargrove and myself were loading some cattle at Albert Neel Huff's when one of the cows started fighting the horse. She run at him several times and he kicked her down a couple of times. Finally, he kicked her so hard it broke her bottom jaw. When he was 8 years old, he collicked. I spent $400 on a Vet, but he died anyway.

I sold one of the other two colts to a team roper, which I had already roped lots of cattle on in the roping pen. He made a good horse. The other I worked to the wagon and plowed the garden with. He foundered later on, and was no good to ride.

- CD

ATTABURY RANCH

Every spring from February to the end of May, I doctored ecoli scours for the Attabury ranch. It was a good time to make rope horses out of colts.

I had at least two colts to start every year, for lots of years. I would have them broke enough to start roping on by scour doctoring time every year. I would carry my scour med and a paint stick to mark the ones I had doctored so that I didn't doctor the same ones over and over again.

I would carry two colts with me and rope a while on one, and then the other. By the time scour season was over, they would sure track and rate and stay back on a rope.

That's the way a good horse is made.

- CD

A PRISON STORY

I worked for the prison for five years as a field guard, from 1963 to 1968. Then, in the summer, you were ready to catch your field force squad out at 15 until 6am. You got back to the gate just at dark. We showered the convicts before we went and ate supper. Every other night was duty night, and you had to stay in the dorm if you didn't live on the prison farm. Your only days off was every other weekend. 12 days on then 2 days off.

I carried No. 3 hoe squad, which was a maximum security squad. I had worked the prison rodeo in Huntsville that year, and they gave me two for one for the five Sundays I worked the rodeo. I put that together with my vacation, and that gave me the whole month of November off.

While I was gone, they gave my squad to Floyd Reed. He didn't know that I had a convict I called Possum, that kept trying to run off. They were working in the woods digging up trees, and clearing land near the river. Floyd was riding an old iron grey horse named Thunder. Old Possum made a run for it toward the river. Floyd grabbed his pistol from his holster and shot at him. When he did, old Thunder jumped out from under Floyd.

On the way down, he pulled the trigger and shot again. Floyd said, "They said I shot at him twice, but I only shot at him once. The last shot was straight up." The dog man was sitting nearby with the dogs. He turned the dogs loose, and they caught Old Possum just as he went over the river bank. It wasn't pretty.

- CD

A BULL CATCHING STORY

A man down at Pennington called me one Saturday night. He told me that he heard that I caught bulls. I said that I did. He asked me if I could come catch one for him. I said, "Well, it will be after church tomorrow before I can get there." He asked, "What do you charge to catch a bull?" I told him $100.00. He said, "Well, I will meet you at the Pennington store and you can follow me." I told him I would be there about 2:30 afternoon.

He met me there and I followed him to where the bull was. He said, "Now I'm going to tell you, four cowboys ran him through nine fences yesterday, and never did catch him." I said, "They don't sound like much cowboy to me."

He had a little four horse trailer he was pulling. He told me he wanted to load the bull on his trailer, and then he asked me what I wanted him to do. I told him when I get him caught, bring the trailer. The bull wasn't very big. Probably a 14 or 15 hundred lb. Brahman bull. He was with a few cows, across a little branch. I sent a brindle dog to bay the cattle, and the bull stayed with the cows. When I drove the cattle back across the branch, the bull left out in front of the cattle. I ran by the cattle and roped the bull, 200 yards ahead of the cows.

The man pulled his trailer down there, and I threaded my rope and pulled the bull in the trailer. He said, "I wish you would ride that horse off in that tank, and at least get him wet. I hate to give you this hundred dollar bill for 8 minutes of work."

But, you never know.... it could have took a week. ~ CD

A FUNNY STORY

We were working cattle for Merle Bobbitt at his Shady Grove place. The chute was pretty short and you could only get about two cows at a time in it. Hester Faye, Merle's wife, was sitting on the tailgate of his pick up. She was about 100ft from the front of the chute, and a little to the left. She had her record book and was writing down tag numbers and ages of the cows.

We turned a cow out of the squeeze chute that was stirred up pretty good, and she headed straight to Hester Faye. She, the cow, hung her horns under the tailgate and closed it. Hester was laying on her back with her legs hanging over the tailgate screaming, "Help me, help me, I can't get up." She couldn't get up until I let the tailgate back down.

She wasn't hurt, but she changed her work place!!! ~ CD

A CATTLE DRIVE STORY

Dr. Jerry Bedix had cattle on both the Frank Smith places, one place was on the west side of Crockett, the other was on the east side. We needed to move some of them from one place, to the other. But, instead of hauling the cattle, we decided to trail them.

Dr. Bedix liked that kind of stuff, and we had trailed cattle several times from the Wade Minter place to the Smith place, but they were on the same side of town. These cattle were gonna have to be trailed over half way around the loop of Crockett. Dr. Bedix got a policeman named Fat Albert to escort us.

This was in 1986, and it was in the middle of the hot summer time.

By the time we got everything set up, and the cattle drove out on the highway, it was probably 10am. When we reached the overpass in front of Willie Josey's Body Shop, it was close to 12pm. We had to keep the cattle inside the guard rails on the overpass or they would fall 50ft onto the railroad tracks.

I was sitting at the end of the guard rail keeping the cattle hazed inside the rail, when a woman drove up beside me and said, "Is this part of the sesquicentennial?" I said, "I don't know much about no sesquicentennial, we are just moving some cattle."

Fat Albert was easing along ahead of the cattle with his police light on, when all of a sudden I saw smoke boil out of the police car. Come to find out, easing along there with the air conditioner on, he had burned his motor up. They had to send another police car to finish the trip.

Glad to say all went well and we got the cattle moved. - CD

WOODS CATTLE

All my life, and my grandpa before, we ran cattle in company woods. The cattle done good in the woods because generation after generation, they were raised there. You can't put pasture cattle in the woods, or they will starve to death.

Every year, I would work my calves a few at a time in the woods. The cattle were well dog broke, and I would bay them up in little bunches, rope the calves, cut and earmark them.

When it came time to sell them, I would tie four or five ropes on my saddle, catch that many calves, tie them off by a foot to a tree, and go back to get my truck and trailer. Then I would load the calves I had tied and do it again. I would get a few each sale day, until I got them all sold.

- CD

I didn't have a pen, and got along fine without one.

A DR. BEDIX BULL CATCHING STORY

Dr. Jerry Bedix had five good Brahman bulls, and one of them wouldn't stay in the pasture. Walter Sherrouse and myself went to catch him one day. I had already roped him out of other people's pastures several times.

This time, he was was by himself on a small place that had 50 or 60 acres in it. It was mostly woods but had one opening with about ten acres in it. We had to park the truck and trailer on the road, because we didn't know where we would find him. The place was pretty deep because it was long and narrow.

I roped the bull on the backside, and drug the bull over to a big Pin Oak tree that had a big limb hanging pretty low. Walter got another rope on him, and I tied my rope off to the big Pin Oak limb. I told Walter to just keep him from tangling up and I would go get the truck and trailer. It was half a mile or so back to the truck, and when I got there, Bob Thompson was sitting in his truck, there by my truck. We talked for a while and I told him I better get on back, that I had left Walter holding a Brahman bull back down there. Bob said, "Oh he will be alright if you've got him tied to a tree." We talked a little longer.

By the time I drove back over to where Walter was, he was fighting mad. He said, "Where in the world have you been? That bull has done jerked my horse down three times." He was pretty upset with me for a while.

- CD

BILLY BOB CHRISTIAN COW CATCHING STORY

Thanksgiving day, Billy Bob Christian and myself went to catch two cows for a man. It had been misting rain all day and the place was red clay that was plenty slick.

I was riding a big blaze faced Sorrell colt, that I had just broke. I had roped a good many cattle on him, but he was pretty green. We found two cows over on an old red clay hill. Billy Bob took to one going up the hill, and I took to the other going down the clay hill. Just as I roped the cow, she made a right turn behind a persimmon tree. The horse kept straight, and when the rope came tight it jerked my horse down on his side.

The first thing that hit the ground was my right elbow. It dislocated my right shoulder. I couldn't even sleep at night it hurt so bad, and all day long too. Marian Willis kept telling me that I was going to have to go to the doctor and get it put back in place. I said, "Oh it will get back in place sooner or later."

The next spring, I was plowing my garden with a horse and a Georgia stock plow. The plow hit a stump and it threw me over the plow handles. It pushed my shoulder back in place. It hurt so bad for just a little while I could barely stand it, but it sure felt better a few weeks later with everything back in place.

- CD

A COW IN THE LAKE STORY

James Ashley and myself was penning cattle for Billy Gibson on the old Driskell place. A cow had swam out in a pretty good sized lake, and started swimming in a circle. I told James he needed to swim out there and run her out of the lake before she drowned. He said, "*I'm* not going to."

So I tied my horse, and stripped off my clothes and swam out there and run her out of the lake. When she came out on the bank, James took to her and roped at her and missed. She made a circle and jumped back in the lake and swam out in the middle again. I told James "I'm gonna run her out one more time, and this time, just keep her from getting back in the lake." When she came out next time, I run, jumped on my horse without a stitch of clothes on, and roped her. James made his rounds for weeks telling everybody in the county about it.

What happens in the pasture, should stay in the pasture.

- CD

A WAGON TRIP STORY

For years my family and I went on vacation in our covered wagon. The trips we enjoyed most were the one's to the Neches River, where we would camp for several days.

Sometimes we would travel at night so it was cooler on the horses, because we usually went around the 4th of July when the weather was very hot. I would back the wagon up near the house and my wife would load it with what we would need while we were gone.

This particular year, we left the house about dark. I unsaddled my horse that I worked that day, harnessed the wagon team up, and we left. By about 3:00 am we were 25 miles away, and pulled the wagon over in Hayes Park. I told my wife I couldn't go any farther, and I tied the horses to a Post Oak limb.

There was a mattress in the bed of the wagon that the kids were sleeping on. My wife stretched out on the spring seat of the wagon and I layed down with the kids.

Before we left, my wife had fried some chicken for the kids to eat on the way. There was a house cat that must have smelled the chicken, snuck his way onto the wagon and joined us on the trip unnoticed, hiding in the fly of the wagon sheet. Just as we dozed off, the cat climbed up onto my wife's chest, and she let out a blood curdling scream!

That cat made about four trips around the wagon before he found his way out. It scared the cat about as much as it did her. The horses were so startled at the scream, they broke the limb they were tied to, but didn't run. When we woke up the next morning, the horses were still tied to the broke limb.

We made it to the river the next day and set up some tents for the kids to sleep in. The two boys wanted to set their tents up across a little branch of water, where they would be away from my wife and I, and our baby boy Matthew.

We were in a big tent sound asleep when all of a sudden it started pouring down rain. Our tent was leaking so bad, that we had to get up and get in the wagon, which was waterproof.

I had fixed the tailgate of the wagon so that my wife could let it down and use it like a table. The next morning, that little branch was as big as a river, from all the water coming off the hills. We had parked the wagon in an old dry slew, and now there was a foot of water all around the wagon.

My wife was wading water fixing breakfast on the tailgate of the wagon. The boys couldn't get back across the branch until the water ran down. We propped our bedding up on some sticks out in the sun, and got it dry before dark.

About dark, it started thundering and lightning again. My wife said, "We are getting out of here right now!" We harnessed the horse, and headed out in the dark. We traveled all night and part of the next day, before we got home.

My wife said, "My next vacation will be at the Holiday Inn!" But, we went on lots more wagon trips. I told my wife, "We are building these kids some memories", and we did. ~ C D

CD Langham, son-Matthew Langham,
granchildren-Taylor (Tsamouris) Poston & Trevor Marsh

A CALHOUN RANCH STORY

The Red Town and Puzzlewood ranch was 14 thousand acres of Trinity River bottom. They ran twenty five hundred cows and whatever calves they weaned until the first week in August. Then we gathered over two thousand head of yearling and shipped them.

Gathering time was a big deal every year. The steers were turned out in Black Land Lake pasture. We gathered them out of the brush. Then we trailed them through Red Town and out on the county road, then into the middle of Puzzlewood ranch where the pens were.

We penned, sorted, and loaded steers one day. Then we penned and loaded heifers the next day. There would always be some mixed back with the cow herds, and we would go from herd to herd and rope the yearlings out of the cow herds.

The day we shipped, there would be 14 big cattle trucks lined up for a mile. We would put 71 or 72 on each truck. The cattle usually weighed around 800lbs. each. The next day we would ship heifers and there would usually be about the same number of heifers. Another 14 truck loads.

In the winter, we would always have lots of sick yearlings to doctor and that would be a busy time, until we got the calves straight. Spring brought calf working. Cutting bull calves, burning horns, black legging and ear marking. Worming and leptoing cows at the same time. Testing and working bulls to put out. There was always something to do.

David Mason was the ranch manager, and Dickie Manvel was a full time hand that lived on the place. We only day worked down there most of the time, but sometimes we worked a week at a time. David would say go get dinner for tomorrow. We would take the dogs, go catch and skin a wild hog, and O.C., the shophand, would put him on the pit. That would be dinner for the next day. We would do that everyday through cow working time.

That was good times. They are gone now. The ranch was sold to separate owners. One owner has the 8000 acre Red Town, and another has Puzzlewood with 6300 acres in it.
Nothing stays the same.

- CD

THE ATTABURY 7A RANCH

The Attabury Ranch was one of the biggest cattle operations in the county. For years, we processed and pre-conditioned enough wheat cattle to fill two sets of feed yards that the Attabury's owned near Amarillo.

They turned the calves out on wheat, after they were preconditioned on a gain program. They paid 35¢ a lb to the wheat farmers for gain. They weighed the cattle on and weighed them off, to determine the gain. Besides the calves they processed, they ran 1500 head of cows.

Bob Thompson was the ranch manager, and the fall of the year, he kept me plenty busy day working. I was either driving up sick cattle to doctor or pasture doctoring sick cattle or loading and unloading trucks.

Day and night we doctored cattle and loaded trucks. Over a four month period, we pre-conditioned lots of cattle to go on wheat. Also, we weaned the calves off of 1500 cows, palpated and wormed and leptoed all the cows and bulls.

Later, Attabury bought 73,000 acres near Amarillo that he did all that on. He built pens and pre-conditioning yards there. They have now sold the ranch in Houston County and moved their operation up there to Amarillo.

Times change, people get old, and all they can do is tell stories. At least I have a few stories to tell. - CD

CD Langham at Attabury 7A

A HORSE BREAKING STORY

While living down on the 7J Ranch, a neighbor to the 7J wanted me to break some horses for him. He had bought two mares that he never did anything with. One of the mares was bred, and she had a horse colt. When the colt got old enough, he bred the two mares and they had colts every year after that. The older horses were 5 years old, down to yearlings.

T.J. Maples, the owner, wanted me to break everything that was old enough. There was 6 two to five year olds. Of course none of them had ever had a human hand on them.

I got them up in the lot and started breaking them. You could just throw your rope on one, and the fight was on. There was four tenant houses there by the lot, and they would line up on the fences to watch me break the horses. I got two of them broke, and a black man, that had been sitting on the fence, wanted to help me ride some of them.

I had an old blue Chevrolet pickup that I wasn't using, and I told him I couldn't pay him, but I'd let him ride that pickup out. He jumped at the chance, and I would let him ride the ones that I already had peeled out. I had a good breaking saddle, a *Jumbo* with 20 in. undercupped swells on it.

One day he said, "Mr. Carroll, I want to break one of them horses." I said, "Which one do you want to break?" He said, "I want to break that red paint filly."

So, I threw a rope on her and put that *Jumbo* saddle on her. I earred her down, and he stepped up on her. I never saw anything like it in my life. He had one hand on the saddle horn, and one hand under the Cheyenne Roll cantle. That mare bucked as hard as I ever saw one buck, but she couldn't throw him.

We had all six of them broke, and were just putting some riding on them, when an old black man named Jake, that had been watching us break the horses came to me. He said, "Mr. T.J. said if you would break that stud horse for me, that he could be my horse." The stud was the sire of the colts we were breaking, he was a nine year old. I told Jake that I didn't know if he could handle that horse or not. Jake was close to 80 years old.

The next day I threw my rope on the stud horse, and he nearly tore everything down before he finally quit fighting the rope. I got my saddle on him and when I stepped on him, the rodeo was on. I thought he would never quit bucking, but he finally gave out. I got him to where I could start and stop him, and pull him around. I told Jake to open the lot gate and let him out in that 150 acres.

When he came out of the lot, he broke into again. He bucked nearly all the way around that 150 acres. The next day, I saddled him again, and it was all to do over. He bucked about as bad as he did the first day. When I got through that second day, Jake said, "Mr. Carroll, that's good enough, I can handle him from here."

He left leading him towards his house. I had to drive by his house on the way down to where I was breaking the horses everyday, and Jake had the stud staked in his yard.

One day, I was down there riding the horses and here came Jake on that stud horse. He had him in a little jig.

I figured that horse would kill him, but I was wrong. ~ CD

A MARVIN CUTLER COW CATCHING STORY

I had a job catching some cattle for a man in Augusta. I got Marvin Cutler to go help me, but first he had to go sell some walnuts. He had picked up 5 feed tubs of walnuts and he had them sold for a dollar a lb.

It was 10am before we got back from the walnut trip, so we got a late start catching. It was on a Wednesday and the sale started at 12 noon. We got everything caught, but one was tied in the woods. I told Marvin to take what we had loaded on to the sale, and I would take my trailer and load the cow we had tied in the woods and bring her up myself.

Marvin had tied her to a tree by a back foot. When I rode up on her, she ran at my horse. The rope came loose from the tree, and she ran under my horse. He was trying to get away from her and he ran into a tree. It almost knocked me off my horse. When my head went down, his came up, and our heads collided.

It busted my head right above my eye. I busted it all the way across and my top eye lid drooped down. My eyeball was looking out the hole above where it was supposed to. I got the cow loaded, and loaded my horse behind her, and started to the sale barn. I was driving down the highway and a man in a little pick pulled up beside me and motioned me over. I thought I had something wrong on my trailer.

I pulled over and stopped, and he came up beside me and said, "Are you alright?" I said, "Yes, I'm alright." He said, "Man, you ain't alright, your eye is looking out the wrong hole. Let me take you to the hospital." I said, "No, I've got to go move some cattle for Wayne Selman." He said, "You can't go move cattle like that! I'll take you to the hospital, and I'll pay for it."

A good samaritan if there ever was one! I thanked him, but I went and moved the cattle.

When I got home, that evening, my wife said, "That won't ever grow back like that." I told her I didn't have the money to get it sewed up. She said, "I'll call Mary Baker, she is an R.N. and can sew it up."

Mary came right on up to the house and looked at it. She said, "I don't have anything to sew it up with." She called Dennis at the nursing home, and he went by the Vets and got a staple gun to staple it up with.

When Dennis got out to the house, he said, "I'm afraid I'll staple your eyeball. My wife ended up taking me to Temple, to the V.A. Hospital. It was after midnight before we got over there.

When we finally saw a doctor, he said, "We don't have anybody qualified to sew that up. We are going to send you to Scott White, and the V.A. will pay for it." We got to Scott White about 2am, when they finally got me sewed up it was 4 am.

I had to be in Tadmor at 7am. to ship calves for Gary Holcomb, and we had five places to pen. We sorted the calves off, weighed them, and loaded them on big trucks before dark.

I was plenty tired when I got home that night. - $^{C}{}_{D}$

A JAMES ASHLEY BULL CATCHING STORY

James Ashley and myself went to catch a bull that belonged to 7A Attabury Ranch. He was in the timberland on the Hickory Creek Ranch. There is a two thousand acres of timber land that is managed on the ranch.

It is mighty big timber, but some of it had been cut. We finally found the bull in a little clear cut. There was a little grass, but the bushes had grown back about head high. I was sitting on one end of the clear cut and the bull didn't see me, and he came walking straight to me.

When he got close enough, I threw my rope, but the bushes held it up enough that the bull walked under it. I spun me out another loop, and he hit a log road. When I got close enough to rope, he was running fast.

Just as my rope settled on him, he turned off the log road, behind a Pine tree. I kept going straight. That rope sounded like a 30/30 went off when it broke! It hit me in the lower back, and the right arm. There was barely enough skin left on that arm to grow back. The rope broke about halfway between the horse and the bull.

James run up and caught the rope that was dragging on the brush and dallied off.

We got the bull, and I've got the scars to prove it. - $^{C}{}_{D}$

TAKING UP BULLS

Ranchers with a calving season leave their bulls out on cows for a limited amount of days. Usually up to 120 days. If they haven't bred by then, when they palpate in the fall, they get the open cows off. Take-up date is usually the first of July or the first of August. The full time ranch hands would drive up what bulls they could drive up, then Bob would send me to catch the rest.

One year, there was two bulls left on the Bitner place, and Rick Hargrove and myself went to catch them. Rick was running a little late that morning and I already had one bull caught and tied to a tree.

When Rick got there, we found the other one in some woods down in a corner. Rick said, "I will go in there afoot, and run him out, and you rope him." I agreed to do that.

Well, it wasn't very far from that patch of woods to another patch. When the bull came out, I had to hurry. I threw a big loop a long ways. The bull nearly went completely through my loop, but it drawed down in his flanks. There ain't never been a rodeo bull bucked any harder than that!

Every time he bucked, he would jerk my horse thirty feet through the air. He was sure hard to ride, but I managed to stay on him. Rick finally got there, and got another rope on him. I was sure ready for him to get him stopped.

If you're tied on when your rope goes on that bull, it's too late for anything but regrets.

~ C_D

AN ALLIGATOR STORY

I had built a road across an old slew going out to the county road from my house. The beavers cut trenches through my road and it caved in. I couldn't get out that way anymore. Someone told me that alligators would run them beavers off, and I knew where there was plenty of alligators, but it was a long ride down Beaver Dam creek.

James Ashley and myself went over there one morning. We carried some big 14/0 fish hooks and a toe-sack. We rode down there until we saw an alligator. Then we hung a hook and line off a willow limb, just above the water, with a piece of bacon on it. We rode until we saw another one and did the same thing. We got several hooks hung out then we waited a while and came back to run our hooks.

The first one had an alligator on it. We caught him and wet the toe-sack. Then we put the alligator in the sack and tied him behind the saddle.

There was a couple of lines we didn't have one on, but the first one we put out had one on it. Before we could get him out, he got off. We baited the hook again and rode off.

Directly, we heard the limb hitting the water again. We fed that alligator all our bacon and he got off every time. We finally gave up and came home with the one we had.

I put him in that old slew and he cleared those beavers out. I built my road back and everything was fine. He was 5ft long the last time I saw him. He disappeared after a couple of years, I figured someone must have killed him. The beavers came back and ruined my road again.

- C_D

A GENE ALLEN STORY

My friend Gene Allen and myself grew up together. I was a cowboy all my life, but he was a hunter and a fisherman. The best that I ever knew. We had lots of fun together.

Gene had a brindle dog named Rex. He would walk 2 miles through the company woods between his house and mine, and be at my house at daylight. We would hit the woods and squirrel hunt all day. Old Rex, his dog, would tree a coon at night. I told Gene that we should saddle a couple of horses and go coon hunting one night.

I had some green horses I was riding for other people, but I had a broke horse for him to ride. At least, I thought he was broke. I got his horse saddled for him, and was saddling a green horse for me. He led his horse around the little barn and I didn't hear him anymore. I went around the little barn to check on him.

He was laying flat of his back. His mouth would open and shut, but he couldn't say nothing. Finally, he said, "Oh shit, I must have jobbed him with my gun barrel when I got on him. He threw me 10ft in the air and I came down on my back, just like I was laying in the bed. I thought I would never breathe again."

A good while had gone by before I talked Gene Allen into helping me ride some horses I was breaking again. One day, I had one that hadn't bucked in two weeks. So, I told him he could ride that one, because he was through bucking. I had a good Jumbo breaking saddle. I put it on his horse. He didn't get far until the horse broke into. He lost both reins and both stirrups, but he got him rode. When the excitement died down, Gene said, "That's the end of my bronc riding." *....And it was.* ~CD

Gene Allen & CD Langham

ANOTHER GENE ALLEN STORY

Gene worked for Dan Pennington building roads, so he learned how to use dynamite. He brought some in with him one day, and told me that we could go to Pedro Creek and blow up some fish. Gene had a '37 model Chevrolet.

We loaded up the dynamite and some toe-sacks to put the fish in. We found a good hole, tied three sticks of dynamite together and put a cap in it. Our wire wasn't very long, so we set down on the bank pretty close.

When we set the dynamite off, it blew water 500ft in the air. We couldn't hear nothing for an hour. We finally got some fish, but we lightened up on the charge.

From then on, every time we got ready to fire the dynamite, I got way out behind a tree and Gene hollered, "If it blows me up, tell my wife what happened."

~ CD

Larry Henderson, Gene Allen, & CD Langham 1959

THE TENNESSEE STUD

Earl Locke had a 4 year old Tennessee Walking stallion. He was raised in the pasture out of a mare he had, and he'd never had a rope on him. Gene Allen was living on the Floyd Davidson place a couple of miles down the road from Earl Locke's place.

I went by Gene Allen's house one day and I had my saddle in the truck because I had been riding some colts over at my place. Gene said, "There is a black stud that was out in the road, he went on down the road, and now is out in Max McLean's hay field." Gene had a mule that he gardened with and a little Welsh horse in his place and the stud horse was against the fence by his mule. I said, "Let's catch him and ride him."

We didn't know whose horse it was because we had never saw him before. I had a couple of calf roping ropes in my truck that was pretty short, so I tied two ropes together to make one long rope. The stud was pretty interested in Gene's mule and Welsh pony, so I eased up close as I could and throwed my rope on him.

He left from there in a hard run dragging the long rope, but he finally came back up to the fence. The end of the rope was right beside a pretty good size Persimmon sprout, so once again, I eased up and this time tied the rope to the Persimmon sprout.

When the stud felt the tug on the rope, he took off and pulled the Persimmon sprout up by the roots. That Persimmon sprout chased that stud horse all over that hay field. When he finally quit running, I eased up and got hold of the rope.

We finally got him wrestled back out in the road and Gen helped me get a saddle on him. I told Gene to hang on to the rope and don't let him run off with me. I stepped up on him and the rodeo was on....he bucked for a while and then he kicked all four feet out to the side and landed flat on his side. After he finally quit bucking I put a halter hitch on his nose and told Gene to let me have him.

In a little while I had him pulling around and starting and stopping pretty good. I was riding him up and down the road and Earl Locke drove up in his old black Chevrolet pickup. He said, "Boys, I see you found my horse, and it looks like you've got him riding too."

He told me to get off his horse and let him have him. He got on his horse and Gene and I followed him back to his place and he brought us back. Gene told me that a few days later, he came saddling down the road on that Tennessee Stud in a fox trot.

Life's been good to Gene Allen and myself. We've sure had our part of fun.

PS. Earl Locke was the superintendent of Latexo school for many years. By the way, he didn't offer to pay me for breaking his horse.

- CD

Latexo Campus Store Circa 1957

A MATT ANDERSON HORSE STORY

Matt told me that a friend called him one day and asked if he had a horse he could use to rope a yearling on. Matt told him "Sure, he is behind my house in the lot." An hour or so later Matt's cell phone rang and the friend asked, "Is there any way you can catch this horse?"

Matt told him, "Sure, but you can't curse him or you'll never catch him." His friend replied, "Well, I guess I'm done then, because I've already called him everything you can call one."

~as told by Matt Anderson

Matt Anderson

A FUNNY SIGHT

Mr. Lee Carpenter called me one night, and said he had a bull with a mineral feeder on his head. I couldn't imagine a mineral feeder on something's head. I told him I would be working pretty close to his place the next day, so I'd be over to see about it.

We was working in Elkhart the next day, so we went by the Charlie Brown place on the way to work and sure enough, there was one of those mineral feeders that has a sail on it that the wind turns to keep the rain out.

The tub part had came out of the feeder, and the iron frame perfectly fit on his head. That bull looked just like a baseball catcher with his face guard on. I spun out a big loop that would fit over the bulls head and the mineral feeder. I lined him out and roped him and the feeder sure didn't seem to slow him down much.

James Ashley heeled him down and I thought I was going to have to get a cutting torch to get that catcher's helmet off that bull, but I finally got it. I put it in Mr. Carpenters truck so nothing else was tempted to put it on it's head.

~ CD

Mr. Lee Carpenter was a good man and my friend. He is gone now.

A PAINFUL BULL CATCHING STORY

Billy Bob Christian, Darrell Gilmore and myself, went to the Carl Cooper bottom to catch four bulls. We had three of them caught and loaded, but was having trouble finding the fourth one. Finally, we found him on a little dry branch that had a strip of timber on each side of it.

Billy and Darrell ran him out on my side. I was riding a big old sorrel horse that would run off with you. I went to the bull and roped him, but the horse run by the bull and just run off. When I came to the end of the rope, it took my dally away from me.

When it came out from around my saddle horn, it nearly tore my thumb off. There was a hole between my thumb and hand, you could stick nearly an inch of your finger down in.

When we got the bull to the truck and loaded him, I asked Darrell if he had any iodine. He said, "Only some 7% cow iodine." I said, "Pour some in this hole." He took the squeeze bottle and filled it full. Everywhere it run down my hand, it took the hide off.

Darrell told Billy, "Old Carroll don't feel pain like other folks." He said, "If you poured that iodine in that hole on me, you would have had to pour it on someone else to catch me."

- CD

A COW CATCHING HORSE STORY

I went to catch a cow for Grady Lake over on Dale Gallant's place. She was on the back side of the property. I got her roped, tied my rein around the rope, and walked back to get my truck and trailer.

When I got to my truck, William Turner was waiting at my truck. We talked for a while, and then I told him I had to go load the cow I left my horse holding. He said, "There ain't no need here he comes with the cow." I looked up and the horse was coming across the pasture pulling the cow.

Mr. Turner was impressed, and I was very surprised.

- CD

A SAD HORSE STORY

Billy Davenport and myself was gathering cattle for Bobby Shoemake. It was big place with lots of woods on it. We got all of the cattle penned, but sixteen head, that we couldn't find. We rode every patch of woods on the place out, and it was getting close to dark.

There was a pretty good sized creek that had one patch of woods cut off, that we hadn't looked in. There was only one good crossing on the creek that the banks wasn't straight up to the heavens.

We rode across the crossing to get on the other side. Then we rode back and forth letting the dogs hunt the cattle. We ended up way down the creek from the crossing. I told Billy that I thought we could cross right here, but it would be pretty rough.

He was riding a bay mare that would go just about anywhere you pointed her. He bailed off in the creek, but the bank on the other side was high and it was real boggy in the bottom.

When he got across, I bailed in about the same place. When my horse lunged to get up the bank, there was a sharp root growing out of the bank, and it went under the flank strap and through his belly. We couldn't figure out why he couldn't get up the bank. I just fell off in the creek and the horse still couldn't get out.

Billy said, "He is hung on a root", but he didn't know it had went through him. We pushed the horse back off in the creek and he finally got out. Then I saw his guts hanging out about a foot. We was about a mile from where the truck and trailer was. Billy said, "I'll run and get the truck and trailer."

The only way in where we were, you had to go back to the highway and back come down another road. I knew that would take a while. I took my boot off and then took my sock off. I pushed his guts back in and stuffed my sock in the hole, but the guts kept pushing back out. Finally I got the sock spread out enough that it kept the guts in.

We took him to George Beeler, the Vet, and he washed him out and sewed him up. He stayed there at the Vet's for a week or so, but he kept filling up with infection.

My wife went down there twice a day and doctored him. She got the doctor to cut a hole under his belly and she would stick a water hose in the hole that the root went in and washed a couple of gallons of corruption out of him twice a day.

The doctor told her, "You might just as well take him home, because you are doing all the work anyway." We brought him home and she did that twice a day for another two months. I carried him to A&M, and they x-rayed him and said he was going to die.

My wife kept doctoring on him, and he finally got well. Dr. George Beeler said if he ever got sick, he wanted her to doctor on him.

Well, my wife and Dr. Beeler both have died, but the horse is still alive. He has a little hernia where the root went through him, but I spaced the girth hobble strap, so you can pull the flank strap real tight, and it holds the hernia in. He is 22 yrs old now....and still alive.

- CD

He is a good one.

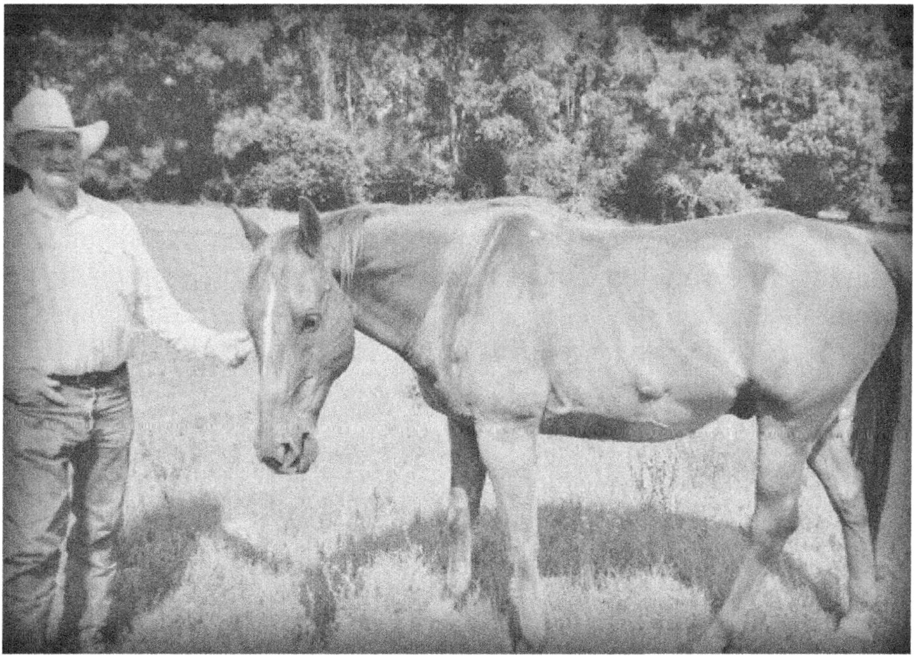

CD & His Horse

A WILD HOG STORY

Day working on the Calhoun ranch, we had moved a set of steers out of the Black Land Lake pasture. We went back to make sure that we had got all of them. It was a big pasture, and it had lots of brush in it.

Dickey Manvel, David Mason, and myself, was riding together. We had two dogs that we were hunting the yearlings with. Both of the dogs would find cattle, but they would also find a hog.

We were riding along the levy, when all of a sudden the dogs ran over the levy and started baying. We tied our horses and went over the levy and saw that they had bayed a hog. Little David Mason picked up a big stick. Dickey asked what he was going to do with the stick. He said he was going to keep that hog off him.

When we got to where the hog was, it was a big old barren sow. She never had any pigs and she was mud fat. Dickey and myself got her by a back leg, and we were holding her. Dickey told little David to get the dogs and hold them. David got the dogs, one in each hand, and got 50 yards out from us to hold the dogs.

Dickey and myself decided it would be a lot of trouble to get her out of there because she was big. We decided to just turn her loose. We turned her loose and she went out there about six foot and turned and came back. She took to Dickey and myself and was running us around and around a tree. I was laughing so hard I could hardly run. I stepped on Dickey's spur and he fell down.

The hog ran right over the top of him. He asked me what the heck is so funny. The sow left out of there headed straight to little David Mason. He threw the two dogs and left out. When he made it over to us, Dickey asked, "What went with your stick?" He said, "That stick was slowing me down!"

- $^{C}{}_{D}$

A BAD WRECK

We were working cattle on the Jr. Minter ranch. It was on a Saturday, and we were palpating and worming cattle. Rick Hargrove and myself was bringing the cattle down the chute and Billy Bob Christian and Mark Avery were working the squeeze chute. There wasn't a gate on the crowd pen, so, me and Rick was the gate.

I would push the cattle up to the mouth of the crowd pen, and Rick would run them down the chute. There was some vines grown up at the corner of the crowd pen, so you couldn't see anything coming out.

Just as I walked up, one of the big tiger striped cows came out of the crowd pen in a hard run. She side kicked with both back feet. It knocked me into the fence, and when her back leg came forward, it went in between my legs. It threw me across the pen and under the whole pen full of cows. They all ran over me when they came out of the corner. I broke my hip and leg in four places.

The cowboys threw me in the bed of a pickup, and hauled me to the hospital. There I layed in the hall of the hospital from 9:30am until 2:30pm before they finally took me to the Tyler Trauma Center. The bones in my leg was broke at such a sharp angle that it ground the muscle in my leg into hamburger meat. I lost half my blood that came out inside the leg. It was five months before I could walk. The leg that was broke is over an inch shorter than the other one.

Good people kept me and my family alive until I could work again.
I thank "God" for these people, which are too many to name. ~ CD

Ranch rodeo to assist Langham's medical bills

A benefit ranch rodeo and barbecue for C.D. Langham is set for Saturday at the Porth Agricultural Arena. Rodeo events will begin at 10 a.m.

Langham was recently injured while working cows. He has had surgery and a lengthy recovery period is expected. Donations will be accepted.

Article from local paper.
For fundraiser/benefit put on for CD Langham

A TRINITY FLOOD STORY

When the Trinity River overflows, it covers lots of bottom land. In 1990, it overflowed. Everybody was getting cattle out of the backwater.

Lee Lundy had a black man working for him that he called Rabbit. Lee and his cowboys were getting cattle out of Clear Lake pasture when the levy broke. They had pulled a trailer across the dragline ditch on the dump road.

While they were gathering the cattle to drive them out of there, the levy broke and the river water was coming down the dragline ditch, and it was full, plus as swift as the river. When they got back to the dragline ditch, they had to swim the cattle and horses. Rabbit couldn't swim and he was afraid of water.

There was a culvert there where they crossed, but you couldn't see the end of it for the running water. Rabbit's horse missed the culvert and the horse was thrashing around trying to get out. Rabbit fell off and the swift water washed him into a tree. He just climbed the tree and stayed there. There wasn't any way to get to him except by boat.

There was still a horse left in the trailer way out in the middle of that big bottom. The only thing sticking out of the water was the horse's head. They took the boat out there and floated the horse back to bank beside the boat.

I had carried my boat in the trailer with me that morning, because I had to get some cattle out from behind the Dooley Hole on the Karo Ranch. They asked me to take my boat and go get Rabbit. But, before I got through getting the Karo cattle, the Game Warden had already went and got him.

Rabbit couldn't swim and he thought the water was deep because it was running so fast. The water wasn't 2ft. Deep, but Rabbit didn't know, and he was afraid of water because he couldn't swim. The fire ants nearly eat him up and he had stayed in that tree 3 hours, but he was alright.

- CD

Trinity River Flood Circa 1990

A BROKE SADDLE STORY

Justin Baker and myself was penning cattle at Bitt Caddenhead's. Bitt told me he had left a Brangus bull in the bottom, and he wanted us to go catch him, load him, and take him to the sale when we took the other cattle.

We went down there, and he had came up on the side of the hill. When he saw us, he headed for the bottom again. He was running along a fence when I roped him. I caught him by the top knot and a front leg. It came off his topknot, but stayed on his front leg. I didn't want to wreck him, so I was trying to slow him down a little at a time.

I had just got my saddle back from putting another tree in it, because I had broke the tree. We were moving along pretty good, when I came to a Black Jack tree. The horse on one side and the bull on the other. The breast collar jerked the D rings through the riggin, and the saddle went with the bull.

Justin was behind him trying to catch up to rope him, when the rope went under a Pin Oak root and stopped the bull. Justin got another rope on him. I jumped on my horse bareback, and went and got the truck and trailer. We loaded the bull and I had to go home and get another saddle.

That's why cowboys are rich.

P.S. My friends had bought me the saddle, and told me I wouldn't break that one. So far I've put two trees and 1 new riggin under it. ~ CD

ANOTHER BROKE SADDLE STORY

Rick Hargrove and myself went to catch a cow on the John Tehotec place. She was a big Brangus cow. There wasn't any way to get the truck and trailer in the pasture, so we left the truck on the county road in front of John's house.

We went in the pasture and I lined the cow out down a fence row and roped her. Rick ran down there and opened a gap to let her out on the road. She tried to turn the wrong way when she came out on the road. I turned the horse toward the truck with her.

She was a big cow, and I was in a lope dragging her toward the truck. All of a sudden, my saddle horn pulled out by the roots. I was tied on and the saddle horn stayed on the rope. Rick was loping along behind the cow with a loop made. He was going to heel her and hold her until I could get my rope threaded through the trailer.

When the saddle horn pulled out, it hit me right in the groin and tore my right britches leg completely off. Then the saddle horn flew by Rick's ear doing about 60mph. If it had hit him between the eyes it would have killed him.

Mrs. Tehotec came running out there and said, "Are you alright?" I said, "I will be in about 6 months." Well, from then on, she knew what I looked like from the hip down, because I was sitting on my horse stripped naked on the right side.

- CD

A BUTCH ARNOLD BULL CATCHING STORY

Butch Arnold, Rick Hargrove, and myself went to catch a bull. We drove up to the place where the bull was, and we were unloading our horses when Butch said, "Do you have a dog that will crawl out of the trailer" I told him that I did. He said, "You had better chain him then, because this man don't want no dogs on his place."

While I was chaining the dog, Butch and Rick rode off. When I got on my horse and started down there, I saw Rick behind the bull, headed toward the woods. I don't know if he ever threw his rope or not, but I saw him jump the fence into the woods.

I ran down to a gate and got in the woods where the bull was, he came out on a fence right of way, and he was leaving here. The fence right of way wasn't very wide, and limbs were hanging out in the right of way.

I finally came to a spot where I could rope, and just as I roped the bull, he went left behind a Post Oak tree. The horse went straight until he came to the end of the rope. The rope jerked my horse over backwards. My head hit the ground so hard that I saw stars for a while.

When the horse got up, my right leg was still in the stirrup. I folded my knee over the saddle, and my right knee hasn't been the same since.

I was laying on the ground when Butch rode up. He said, "Are you hurt?" I said,"My right knee hurts bad." He asked, "Well, why are you holding your head?" **~as told by Butch Arnold**

Butch Arnold in Arbor TX

A DRAGGING ON THE SPLIT CIRCLE

We were gathering cattle on the split circle. We had most of the cattle in a trap, but were missing a few. Alfredo Fajardo had found two cows in some button willows in a lake. When I came along, he had one of the cows roped, and was headed toward the trap with her.

He hollered and told me there was another one in the lake. I was riding a two year old Grullo mare, but had worked on her a lot. My rope wasn't tied on, but I had made me a loop, and hung loop, coils and all, over the saddle horn. Where I rode off in the lake, I didn't know it was a boat landing. When she stepped off the bank it was straight down. The mare was trying to climb back out of the lake, but it was straight up. I just rolled off in the lake so the mare could get out.

My spur picked up the loop when I rolled off and the mare jumped up the bank and took off. The coils drawed down on the saddle horn and with the rubber on the horn, it was like an 8 wrap dally.

I had my leggins on and my knife was in my britches pocket. The mare was dragging me 40mph. I was headed toward a cattle guard and a gate that was open. If she jumped the cattle guard I was dead. If she went around through the gate, I would hang on the gate post.

Alfredo was coming back to help me and saw what was happening. He cut across and roped the mare, just moments before she got to the cattle guard. I was so proud of Alfredo that I wanted to kiss him.

Thanks Alfredo.

~ CD

Alfredo Fajardo

BROKEN SADDLE

Through the years, I broke the tree in 12 saddles. The last tree I broke, my friends began to feel sorry for me. They got together and bought me a saddle they said I couldn't break.

So far I have broken the tree twice, tore the riggin out from under it once, and have pulled three saddle horns out by the roots, two of the stories are in this book. The other wasn't worth telling.

I'm always thankful for friends, and hoping I don't leave anyone out, those that bought me the saddle was:

Dr. George Beeler
Bro. Otis Byers
Rick Hargrove
Ed Anders
Brenda Anders
Perry Jones
Bitt Caddenhead
Bob Thompson
John Olan Turner
Joe Robinson

- CD

CD Langham

A WILD COW TALE

Doyle Oldham had moved all of his cattle from his place near "A" Lake. One Brahman cow was in "A" Lake pasture when he moved the other cattle. He wanted me to go catch the cow for him, and he went with me. He was riding a pretty nice little paint horse. I had a black and tan cur dog, that was a real good find dog.

"A" Lake pasture had 2300 acres in it and only one cow. We took old Joe, the dog, and rode all over "A" Lake before the dog ever found the cow. The old place was pretty rough, with hardly any opening on it. It was all growed up in Ash bushes about 20ft tall.

Doyle and I split up to see if we could hear old Joe bayed anywhere. Directly I rode up on Doyle's paint horse tied on a trail. I heard Doyle saying,, "Help me! Help me!"

I tied my horse beside his, and ran in the Ash bushes to see what was wrong. The cow had a horn on each side of Doyle and he had his arms around her neck. She had pushed him up into some vines as far as he could go. I ran in, and grabbed the cow by the horns and under her jaw bulldog style. Then I told Doyle to get out of here. Then I got to thinking what am I gonna do with this cow? *A good time to start thinking!*

I told Doyle to get the rope off my saddle and bring it to me. He brought me the rope and I told him to put it on her horns and then go around one of those Ash bushes right in front of her, and then go out there and go around another one of the Ash bushes and tie the rope off. He did and I ran the other way. I got my horse and rode in there and turned the cow loose. She left again with the dog on her. She finally circled the whole place until she came back to "A" lake. There was an opening around "A" Lake that was big enough to catch a cow in. When she hit the opening, I roped her. Doyle went all the way back to his place and got my truck and trailer. He had to drive 7 or 8 miles to get where I was, but we got her loaded. Doyle told everybody that I saved his life that day.

Doyle passed away, a long time ago, with Lou Gehrig's disease. It was a lot worse way to go than getting killed by a cow, but you can't choose your fate. Doyle was my friend.

~ CD

A GOOD MAN STORY

Bob Thompson and myself were working cattle over on Beaver Dam Creek. He had rode over there with me, and it came dinner time. It was a long way to go to get anything to eat. There was a little store about 10 miles down the road that cooked burgers, so we loaded our horses and dogs, and took off down there. We were 3 or 4 miles from the store when we came upon a fellow walking down the highway with a duffel bag on his shoulder. He was dressed pretty western with an old bent up western straw hat on. Bob told me to stop and pick him up.

We stopped, the man threw his duffel bag in the back of the truck and started to get in the back himself. Bob motioned and told him to get up here with us. When he did, Bob asked him if he minded if we prayed with him. He said he would like for us to, so we did. By then we were at the little store. Bob asked him to eat dinner with us, but he wouldn't. Then he asked the man if he had any money. He told Bob that he had a couple of dollars. Bob took out his billfold and saw he had ten dollars. He gave the ten dollars to the man, and bid him safe travels.

When we got in the store to eat dinner, Bob said to me, "Carroll, you are going to have to buy my dinner, because I gave that man all my money."

Bob Thompson is a good Christian man, and he is my friend. - C_D

Bob Thompson

A WES MATTHEWS STORY

Wes Matthews was way out in the country, at this little country store. On the front porch of the store was a domino table. Several old men sat on the porch nearly every day and played dominoes.

Across the road from the store was a corn patch. The corn was already tasselled, but there was two teen age boys running plows down the middles. The only time you could see them was when they plowed out to the end of the rows to turn the mule and Georgia stock around.

One of the old men playing dominoes was their daddy. Wes heard another of the old men ask their daddy, "George, why have you got those boys plowing that corn? You know that corn is already made." He answered, "Yeah, but them boys ain't!"

Wes Matthews was a good man. I worked a many a day for him. He always wanted me to go with him so it wouldn't cost him a trip. We would load our horses and my dogs in his trailer. Most of his pastures was in Leon Co., and it was a pretty long haul. He talked real low and always had a cigar in his mouth. He would tell me stories all the way over there, and I would have to strain my ears to hear him. He worked for Rex Cauble on his horse ranch, and most of the stories was about that.

Wes had a stroke when he was fifty-something, and was in a wheelchair the rest of his life. In memory of Wes Matthews. ~ CD

Latexo General Store Circa 1969

MY LAST BIG ROUND UP

Out near Fort Stockton Texas, there's a ranch that has a little over eleven thousand acres in it. It's the smallest ranch around there. A man named Mike Landry lives out in the middle of the ranch with his wife Gale. It is 10 miles to the nearest highway from his little cabin. He is an artist and a painter. His art is all western paintings, I took pictures of them and they are mighty pretty.

Rick Hargrove and myself went out there and spent four days gathering his cattle for him. It is rough old country. Just rocks, Soto, Yucca, Sage, Mesquite and Juniper. Also some Dog Pear and Prickly, plus some more things with thorns.

Rick and I hauled four horses out there, because it was lots of riding. The wind blows all the time and sun will burn a hole in you. No wonder all the people out there have skin like leather.

There are four wells on the place, and the water is gravity fed to the troughs below. There is no water except what comes out of the wells. Rick and myself pushed cattle to a 650 acre trap for three days, then we penned out of the trap and sorted off the yearlings and loaded them on a cattle truck, the fourth day. The closest livestock sale was Pearsal. That was 280 miles away. This is rough country. Hard on animals and people.

The following poem was written about this country. It made an everlasting impression on me. So did Mike and Gale Landry. ~ CD

Mike & Gayle Landry

BAD LANDS

Riding these hills, so wild, and so free
I wonder how much beauty in this world there can be
But how much loneliness can one man stand
Out in this deserted and barren land

The stars are so many they blanket the sky
The valley's so deep, and the hills are so high
While I'm doing this wondering, I can't help but ask why
This place that's so pretty, could be so darned dry

This country's so rough, of which I speak
It's surely not for, a man that is weak
For among this rock, not much will grow
About all that you see, is the yucca and soto

If I never go back to this country again
There is one thing on which you can depend
I will always have these sweet memories
Of the prickly pear cactus and Mesquite trees

~ C_D

Rick Hargrove and myself were riding the hills near Fort Stockton, mule deer hunting, when I wrote this poem.

~ C_D

A BILLY CROWSON STORY

Billy Crowson is the oldest custom cowboy in Houston County. He has worked cattle for the public most of his life. He has got many stories to tell be he won't write one for me. So I'll tell a story for him.

Billy had lots of pardners through the years, but the one that worked with him the longest, was Butch Arnold. They were driving down the road in Billy's truck, and there was a pair of forceps laying on the dash. They are made like scissors, but they lock when you clamp them together.

Billy was driving and Butch picked up the forceps and clamped them on the meat by Billy's titty. There was also a Hauptner gun on Billy's dash. Billy picked up the gun with a ¾ in 16 gauge needle in it, and went to sticking it in Butch's leg. Billy told him to get the forceps off his titty. Butch told him, "Okay, okay, give me a chance. If you'll stop jobbing me with that needle."

They pulled many a trick on each other over the years, but that was a bad one.

- CD

A BUTCH ARNOLD STORY

Butch was feeding in the White Rock Creek bottom when he got his truck stuck. It was only a couple of miles to his house if he walked straight across. He walked back to the house and got his tractor. He was going to take his tractor over there and pull his truck out and drive it back, and just leave the tractor over there.

While he was gone, James Lloyd Murray and Dean Cook came along and saw that Butch was stuck. They went out there and pulled Butch's truck out on dry land, then they went on down the road. When Butch got there, the truck was already unstuck, so he takes his tractor back to the house.

Again, while he was gone, James Lloyd and Dean came back. They went out there and pushed his truck back in the mud hole. When Butch walked back to get his truck, it was stuck again. He was sure mad.

- CD

ANOTHER BUTCH ARNOLD STORY

Butch and myself was working cattle for Willie Josey. I was working the squeeze chute, and about the second cow I caught, the squeeze handle broke off. I threw it down on the ground.

Butch was fly tagging the cows. When he went to punch the tag through the cows ear, she threw her head around and mashed his hand against the chute. He grabbed the squeeze handle on the ground and hit the cow on the head with it. It glanced off her head and hit him on the knee.

After much pain, he told me that if he picked that handle up again, to take it away from him.

- CD

AN OTTIS BOWERS AND TERRY FRITZ STORY

We penned cattle on the Bitner place for Ottis Bowers. He drove back to the back of the place and found a Brangus bull that we had missed. He asked me if I would go back there and rope him, and pull him in the trailer for him.

Terry Fritz was there with Ottis and I told him to bring my truck and trailer. He drove it back to where the bull was and I roped the bull and pulled him in the trailer.

I told Terry to drive my truck back and I would ride my horse back. He told me to load my horse and drive my own truck back. He got in the back of the truck and sat on the tool box because him and Ottis are both big men.

When I was driving back, Ottis told me there was a wet spot that I needed to get a little run at. Well, I got a big run at it, and it was rough. It threw Terry off the toolbox and almost out of the back of the truck. He has called me A.J. Foyt ever since. He also told me he would never ride with me again.

Bro. Ottis Bowers has been my preacher, and also my friend, for many years. Also Terry Fritz was my friend up until I almost threw him out of the truck, I hope he still is my friend.

- CD

A PARADE STORY

Darlene Brown had a western wear store. She wanted me to represent her western wear in the Crockett parade. I was working cattle for Skinner Brown the day of the parade, so I drove my wagon up to the store.

She and her daughter Darla, decorated the wagon while I was working cattle that day. They ran battery operated Christmas lights all down the britchen, and around the hames and collar.

We loaded up on the wagon. Darlene, Darla, her baby and myself. We was the last entry in the parade. What I didn't know, was that we were going to start at the First Baptist Church, go up to the square, around the square, and back to the First Baptist Church.

My wagon team had never been to town. Skinner was out riding on old Goose, a big old long necked bay horse, ahead of us. Way on up in the front of the parade was the Crockett High School band. They went up and went around the square and came back meeting us about the post office. They were blowing them horns, beating on their drums and stomping their feet in rhythm.

I told Darlene & Darla to get the baby and get off the wagon, that we was fixing to have a runaway. They jumped off the wagon, and took off. The wagon team was so scared that they couldn't run. They just stood there and trembled.

That was the last parade for my wagon. ~ CD

CD Langham - Crockett TX Parade

BULL IN THE BUS STATION

A DR HARKIN'S STORY

Houston County had a Vet by the name of Dr. Harkins. He practice here for close to forty years. He was real layed back, you never saw him excited about nothing, and he was really funny.

One day, Thomas McLendon hauled a calf in to Dr. Harkins' to get it treated and he was pulling a little 16x5 Gooseneck trailer. He had the back gate tied with a rope as extra security because he didn't trust the latch. Dr. Harkins asked him, "Why do you have this rope? Do you not trust your latch?" Thomas said, "Well, I'm not too sure about it." Dr. Harkins asked him if he wanted to sell the trailer. Thomas said that he would. Dr. Harkins said, "My son-in-law needs this trailer," so, Thomas sold Mr. Rains, Mr Harkins' son-in-law, the trailer. Mr. Rains took the trailer and completely redone it. He sand blasted and painted it, but didn't do anything to the latch.

Mr. Rains brought 4 Limousin bulls down to the Vet clinic to get them fertility tested before he put them out. After Dr. Harkins got them tested, he loaded them back on the trailer and started home with them. When he started to pull out on the highway the gate came open and all four bulls unloaded in the highway. Well, these bulls was pretty stirred up from what they had been through. Needless to say, they left there in a hurry. Three of them turned right at the intersection and one went straight.

Down the road a little ways was the bus station. The door was never closed if anyone was there, so the bull went in. The attendant was an elderly woman sitting at the desk, and here came this runaway bull right on in. She jumped on top of her desk and went to screaming, it scared the bull so bad he jumped through the 10x12 plate glass window! They was going to file suit on Dr. Harkins for damages. He said, "Well, I may can afford the plate glass window, but I don't know what the heart attack is going to cost me."

When someone brought an animal in that Dr. Harkins' couldn't fix, he would say "I wouldn't spend no money on this animal, I'd carry it to Vernon Calhoun's kill plant in Palestine."

Dr. Harkins is gone now, but he was one of my favorite people.
We miss him. ***~Thomas McLendon story written by*** ~ CD

A SCARY BULL STORY

Dr. Jerry Bedix had bought three trailer truck loads of Hereford cows from out in west Texas. That was in 1981, they had a bad drought that year, and there wasn't any grass growing out there. He didn't give much for the cows, and he had bought the old Wade Minter place which hadn't been used in a good while.

There was some grass on it, so he put the cattle there. Some of the cows was bred when he bought them even though they were poor. He went up to Bill Huff, a registered Brahman breeder, and bought six good Brahman bulls. He was going to breed the Hereford cows to the Brahman bulls and get him some ½ Brahman heifers.

There was no pen on the place, because it had been a cotton farm up until then. Down the road a couple of miles was another place he had leased that belonged to Frank Smith. Dr. Bedix wanted me to trail the cattle from Minter place to the Frank Smith place. I gathered up a couple of extra hands, he called the Highway Patrol office, and they sent an officer out there to escort us.

We bayed on the cattle a good while to get them bunched good so they would drive good and stay bunched. The Highway Patrol was holding up traffic so we could get the cattle out on the highway. The front car that was sitting still until we got them out on the highway was a little red foreign car that was real low on the ground. When we drove the cattle out on the highway, all the Brahman bulls but one, was behind the cows and calves.

As we're trailing along, all the bulls but one went with the cows. The biggest of all the Brahman bulls walked up to the little red car and stuck his big head right up to the windshield. That woman screamed and hit the floorboard. She didn't stick her head back up until we was way down the road.

Dr. Bedix has moved back to Florida, and Bill Huff has passed on, but his sons and grandsons are still in the cow business. Although they're not as big in the Brahman business as they once were.

Time moves on, things change, but I've still got my memories. ~ CD

A DOG NAMED SAM

I had a big Airedale dog named Sam. He was pretty famous around Crockett Texas. Everybody knew him and he got his picture in the paper.

Houston County Livestock wanted the Gillette Bros. to represent them in the Christmas Parade. They had their Chuckwagon and grey mule wagon team. James Jackson and myself was out riding for the Gillette Bros. and we was dressed pretty western with our leggins, and James had on big rowelled spurs and jingle bobs.

Sam, the Airedale dog, was right alongside my horse. That's when he got his picture in the paper. Merle Bobbitt picked him up in town one time, and took him to George Beeler's and put him in a cage. Ol Sam thought he was in jail, when I went and got him. He was a sad dog.

~ C_D

A SNAKE STORY

My family and I went to the river in the wagon, every year for several days. It would always be around the 4th of July, when work was slow. James Jackson was working for me at that time. He didn't have anything to do while I was on the river, so I told him to bring me my boat and motor and we should do some fishing.

About the second day we was down there, James showed up and he had his brothers Billy and Johnny with him. We put the boat in, and went up the river fishing.

When we were coming back, there was a big fish snake laying on a stump. Billy motioned toward the snake, and picked up the boat paddle. I turned toward the snake. He scooped the snake up on the boat paddle, and flipped him in the boat. Billy and James just sat there, but Johnny jumped in the river. The snake crawled out of the boat, but Johnny was plenty mad.

When we got back to camp, Johnny told my wife, "My brother's are crazy."

~ C_D

GERONIMO!

Before I begin this story, it's probably best that I try and give it a little context. My brother-in-law, Randy Allee, had a modest heard of mixed high quality commercial cattle and a couple of bulls to service his heifers. One would never know, judging the man from the outside, that he was quite a tender-hearted soul, but I'm here to tell you that he is. As he built his herd over time, buying heifers and the bulls he named each- one of the bulls being given the name "Geronimo".

Some years passed after my brother-in-law began dabbling in cattle, he, my brother Jimmy Neel, and I, came together and decided that we would go in together and buy a small herd of cows. This herd would reside on the property where I lived, out near Augusta TX.

We began buying heifers first, some open (not bred), some bred, and some cow-call pairs. After our herd had reached the size we wanted, we began looking for bulls. My grandfather on my father's side, Howard Neel, had spent over half his life in the cattle business and a large number of his herd were Hereford cows.

So the three of us, being a bit sentimental, decided that we would get two Hereford bulls to put on our herd. So we found them, paid for them, but they were still a little young to begin their job, so it would be a couple of months or so before they would be ready to come home.

Now my brother-in-law being far more experienced than my brother and I, was well aware that having open cows remain open for at least two more months, was not a good thing and not an efficient start to our little operation.

On a crisp, beautiful Saturday morning, at about 7 am, I'm sitting in my recliner finishing my 2nd cup of coffee when I hear the unmistakable sound of metal banging loudly, from what sounded like the north side of our property. I've heard cattle trailers pull into this particular entrance of our property adjacent to our house before, but this sound was different....much louder, and much more "violent" sounding.

I sprung up, tip-toed into the bedroom where my wife was still sleeping, threw on a pair of jeans and slipped outside. I went out the front door and immediately looked toward the side entry to our property and saw Carroll's truck and a long paneled cattle trailer.

The sounds coming from that trailer were something from a horror movie- iron clanging and squeaking, loud thuds and the sounds of an animal snorting. As I walked briskly over to the truck and trailer, Carroll has now exited his truck and is standing beside the trailer.

I walked up to him, shook his hand, and asked him, "So, what's going on Brother Carroll?" He explained to me that my brother-in-law had called him the night before and asked him to come to his place, pick up Geronimo, take him to Dr. George Beeler's place (a Veterinarian) have him polled (horns cut off) and then take Geronimo to Augusta.

As he was explaining this to me, I was taking stock of this beast in this cattle trailer. This bull, a very large- probably close to a ton in weight, was throwing himself from one side of the trailer to the other, banging his body and sometimes his head on the sides of the trailer. He was undeniably unhappy and discontent.

Blood was dripping off the two areas where his horns once grew, running down each side of his face and mixing with the slobber and drool coming from his mouth and making long strings that were flinging around and jiggling in the air as he went from side to side in this trailer.

He was snorting and with blood shot eyes, making it well known that he was unsettled and angry, for having been uprooted from his harem of "girls", his horns removed, and now taken to a strange place.

As I'm taking all of this in, or doing the best I can to take it in. Carroll removed his cowboy hat, and scratched his head saying, "You know, when he first told me about this job, it didn't sound like a big deal, but now it looks like this may not be a good idea."

I walked around to the front of the truck, opened the gate and he pulled through. I walked ahead and opened a side gate which would give him access to the barn with an attached corral.

Carroll expertly pulled his truck and trailer through this gate, and in very short order, had backed up to the gate of the corral. He exited the truck and we began to discuss the next move. "Greg, I reckon we should turn him out in this lot here and let him get used to being up here, before we open the gate and turn him out with the cows," Carroll said.

I looked around, surveying the corral and the beast in the trailer, and said something to the effect of, "that's definitely the best plan, but, I'm not so sure that this corral is going to hold him Brother Carroll." We are both laughing now at the situation we found ourselves in.

Geronimo was still not amused. Having no other options, Carroll went around to the back of the trailer, and quickly and carefully, swung open the gates to the trailer. Geronimo barreled out of the trailer in a flash. Without missing a step, and in a flash, he runs to the back of the corral, and attempts to jump over the half boarded, half hog paneled 6ft fence of this corral. Needless to say, he didn't make it over, but as it turned out, that mattered little. The fence of this corral was no match to the sheer weight and force behind this attempted leap of this mad bull, and it gave way of course.

Geronimo, I kid you not, never missed a beat, and never lost momentum. He is now in a full trot. About 150 ft directly in front of Geronimo, is a standard 5 wire barbed wire fence, which he goes right through like it was made of we spaghetti. He never changed his pace, and he never lost his speed.

He then continued through the trot over the crown of this roughly 20 acre hay meadow, and we lose sight of him. Carroll looked for him, on horseback and on foot, week in and week out, but no Geronimo. All he found were sagging fences where he had obviously made his mad exit off our place. We put a notice on the local radio station, KIVY, week after week, but no reported sightings of Geronimo came in.

Roughly 4 months later, while working on a farm two places northeast of ours, Carroll spots a large bull, laying in an opening, surrounded by woods. He approaches carefully and once closer, is able to make out that it is Geronimo! He had been laid up in the woods, coming out to graze on grass only when he was hungry, then retreating back into the forest, never "taking up" with the heifers on that ranch.

Carroll, with the help of another hand, was able to rope poor old Geronimo and take him back home, not to Augusta where his services were no longer needed, but back to his "original" home at my brother-in-law's place. Where he died of old age, lots of years later on Randy Allee's place.

~Story by Greg Neel

Geronimo

Greg Neel's Grandfather Howard-
Pulling bale of cotton with horse & steer.
The odd team.

A ROPING STORY

Several times in my life, I've put on ropings. First calf ropings and then team ropings. When Bob Thompson came to Attabury Ranch, he brought a steer chute with him. He told me we needed to put on a team roping. I told him I would buy us some roping cattle.

So we set up one of the biggest cattle pens we had as a roping pen. We put on a roping every Saturday evening for a long time. We had lots of roper's that learned to rope there. Some of them are still roping today. Most of them have got too old. Darrell Gilmore and myself put on ropings on Friday night in Grapeland for a long time. Then we moved to Crockett and put on ropings for a while.

Rick Hargrove and myself bought forty steer and put on ropings for a good while all over the country. My son's brought the cattle and worked the release gate. We had lots of fun, but it sure gets old.

Staying out all night, is not my game. I like going to bed early. - CD

A COLORADO STORY

Rick Hargrove had a hunting outfit up in Colorado. When he came back, he went to helping me work cattle. All summer he kept telling me that he knew where all the deer and elk was. He said we should take some riding horses and pack horses and pack up the Wolf Creek Pass Divide trail about 14 miles and kill an elk next fall.

We got two others to go with us, Gerald Alexandra and Gene Allen. We unloaded our horses on top of Wolf Creek Pass, and packed in fourteen miles up the Continental Divide. You can't pack near enough bedding to stay warm up that high. We almost froze to death, but we killed an elk.

Then we had to pack it back down, plus all the other stuff we had packed up there. I'm not sure that we were having fun. We packed back down a ways and set up camp at Fawn Gulch. We all killed a mule deer buck. Gene Allen kept worrying about his horse running away and leaving him, so before he crawled in his tent every night, he would shine his flashlight to make sure his horse was still there.

One night, we had dressed three deer, and threw the trimmings a hundred yards from camp. Gene shined his light to check on his horse and he told me, "Come look at this mountain lion." I was already in my tent, and I said, "There ain't no mountain lion this close to camp." He told me, "I said, come look at this mountain lion!"

I crawled out of my tent, and we both shined our flash lights on him. He was lying flat on his belly near the deer trimmings. He got nervous and crawled down a little ditch until he got out of sight. The next morning, he had came back and got the deer trimmings. We went back to Colorado many more years, and always killed lots of game. Most of the time we had to carry someone different, because one trip will usually do most folks in. I'm too old for that now, but I have the memories.

Greg Beavers, Henry Beaver, John Turner, and myself went two years ago. It snowed sideways all the time we were up there. It blew our tents down and everything we had got wet. Our horses nearly froze to death and so did we.

That will be my last hunting trip to Colorado. - CD

CD Langham, Rick Hargrove, Gene Allen
(and Gerald Alexandra behind the camera)

A CALHOUN RANCH STORY

The Carter Ranch is directly across the Trinity River from the Calhoun Ranch. Three 2 year old steers had swam the river and got on the Puzzlewood division of the Calhoun Ranch. It was spring gathering and we were working the cows on Puzzlewood.

The set of cows we had penned that day, had the three steers in them. David Mason was the ranch manager. I was bringing the cattle horseback. David told me to sort the three steers off, and put them in the corner pen, because the Carter Ranch was sending a truck to pick up the steers.

I did, but the big brindle steer went around the pen one time, and jumped out. He hit the turnrow headed toward the river, and jumped in with the last set of cows. David told Cliff Pyborn, Bill Merchant, Billy Christian, and myself to go catch him, because the truck was on it's way.

We loaded our horses on a trailer, and took off. Cliff, Bill, and myself were going to rope him and Billy Bob was going to bring the trailer when we got him caught. Bill rode up beside me and said, "Carroll, what ever you do, don't let Cliff rope that steer or we'll never hear the end of it."

When we got within about 300 yards of the cattle, Cliff already had the steer spotted in the cattle. He busted out across there ahead of me and Bill. He went in the cows wide open, and pushed the steer out the other side of the cattle. Bill and I was right behind him. Cliff ran up behind the steer and was just fixing to rope. Bill was trying to make him miss, and was yelling, "Look out! Look out!"

About that time, Cliff's horse stumbled. (Cliff had a big dip of snuff in his bottom lip.) Cliff went off over the horse's head and the first thing that hit the ground was Cliff's nose and bottom lip. It rolled that dip of snuff out of his lip and filled it up with black land. It sure did skin his nose too.

We caught the steer, Billy Bob brought the trailer, and we loaded him. But Cliff sure didn't get to rope him. Carter Ranch came and picked up the steers. We had lots of fun on the Calhoun Ranch, but it sure didn't pay much.

We was fussing about the pay one day, and David Mason said, "You know, there's a ranch in Idaho that charges $200 a day to get to be a cowboy for a day. I'll just get me some of them. Then if their horse bucks them off, I'll charge them another $200."

- CD

A DOG STORY

Ronnie Goolsby and myself was hog hunting on the Neches River. We wasn't having much luck on this side of the river, so we decided to swim across the river to the other side.

So, we swam across, but it was so boggy our horses bogged down when we came out. We had to get off the horses so they could get out of the mud. Ronnie's horse came out on the right side, but mine swam back across the river. I had to swim back and get my horse, then swim the river again.

When we started to swim back across the river we decided to find a better place to cross. We did, but the bank where we came down was straight down. About half way down was a ledge, about five feet from the water.

All the dogs swam across with us except the one that was off chasing a deer. When we got back to the truck, we waited and waited for the other dog. He never did show up. Ronnie don't hear very good and I don't either. I told Ronnie that I thought I heard the dog bark. It was probably a mile or so back to the river, and it was almost dark.

When we got back to the river, the dog was hung up on that ledge above the water. Ronnie tried to call him back across the river, but he wouldn't come. Ronnie finally stripped his clothes off, swam across the river, reached up and grabbed him by the foot and jerked him off in the river. He swam on back then.

We decided what we heard wasn't a dog, but a hoot owl, hooting. We just happened to go back to where we crossed.

It's been a heck of a ride. - CD

A LIGHTNING STORY

James Ashley and myself was in the Cindy Garner bottom by Pedro Creek. It came up the worst lightning storm I have ever seen! There was a big deer stand out in the middle of the opening. There was only one tree. It was a big Pine tree about fifty feet away from the deer stand.

It was raining so hard you couldn't see. James was riding a blue roan saddle horse, that belonged to Ed Kane. We rode over to the deer stand, and climbed up in it. We held our reins through the holes you shoot out of. The lightning was flashing solid. I told James, "You know, with all this iron I've got in me, I'm just like a lightning rod." He said, "Don't talk like that, I'm scared to death of lightning."

About that time, lightning hit that big Pine tree behind us, and the air just filled up with fire. Pine straw rained down all over the deer stand. James bailed out of that deer stand, and on his horse and left like a streak.

It finally quit raining and I got out of the deer stand and started riding back to the truck. The truck was about a mile down an old log road through the woods. I could see his horse tracks in the sand and they was scooping out dirt, telling me that horse was running very hard.

When I got back to the truck, James had his horse in the trailer, and he was in the truck. Both of them had done dripped dry. I told James I was writing a book. He said, "I can fill up a book just with the five years we've worked together.

~ CD

James Ashley

A CALEB RHONE STORY

Caleb started helping me when he was still in school. His Grandpa told me that he hoped I didn't let Caleb down, because he wanted to do the same thing that I did. I don't guess that I let him down, because he is living his dream. He has got a pretty good business. I had lots of fun working with him, because he is fun to be around.

I rode an old bay horse, that was bad to buck when you first got on him. It was a bad old cold day, and I told Caleb that he might have to get on him for me that morning. He said he would if he had to, but he sure didn't want to. Caleb made two bronc rides while he was with me. One jump one time, and one and one half the second time.

He is married now, to a lady that's way too pretty for him, with two sweet babies of his own. She got a good one, and so did he. The 2011 drought took lots of cattle out of this county, but he has still got plenty of work.

How many people get to live their dream? Good luck Caleb. - CD

Caleb Rhone

SIAMESE TWINS IN THE BIG THICKETT

Mr. Norman Nelson had four hundred head of cattle down at Sour Lake. There wasn't any cowboys in that country, because his was the only cattle down there. His place was surrounded by the Big Thickett Wilderness Area. We would go and stay two days because that's how long it would take to work the cattle. He had a trailer house we would stay in, and he would cook for us. He had 200 head of King Ranch Santa Gertrudis cows that still had the King Ranch brand on them, the rest he raised out of the Santa Gertrudis cows.

Mr. Norman Nelson called me from Sour Lake one night. He said he had a cow trying to calve, but couldn't. He wanted me to come the next morning and catch her and pull the calf. I went down the next morning. It is one hundred forty miles to his ranch from my house. The cattle was in the back pasture of 1200 acres. He drove his truck and I rode my horse to where the set of cattle were. I roped the cow and got off with my heel rope, then roped her by one leg, so I wouldn't pull her down.

Mr. Nelson backed up to me and I stuck the horn knot down over the ball on the back of the truck. When I started feeling around, I found four legs and two heads. I thought maybe she was trying to have twins both at the same time.

When I got them pulled, they were Siamese twin bull calves. They were grown together, belly to belly. They were already dead, but they hadn't been dead long. They couldn't have lived anyway, but it was unusual. Mr. Nelson took pictures of them, with me and a Mexican boy holding them between us. They wasn't much heavier than one big calf. I still have the pictures.

Mr. Nelsons passed away but he kept in touch until he died. He sold out of the cow business long before he died, but he was a loyal customer. Most all my old customers has either died or went out of the cow business.

~ CD

CD Langham & Ranchhand with Siamese twin calf.
At Norman Nelson Ranch in Sour Lake

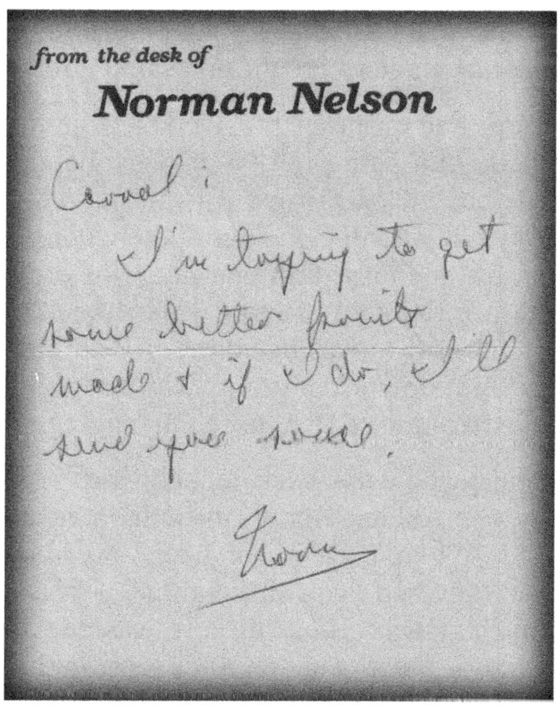

from the desk of

Norman Nelson

Conrad:

I'm trying to get some better prints made & if I do, I'll send you some.

Norm

A SKINNER BROWN STORY

Working cattle on the old Bill Maxwell place, between Lovelady and Weldon, at what we called the Old John pasture, there wasn't a very good place to pen. It was across a dirt road, and was mostly woods with no traps or cross fences. There was just a little opening around the pen.

There was an old black cow that always got away when we penned. Skinner told me that he wanted me to catch that cow. "She needs to go to the sale. She is going to cause everything else to run off," he said. Sure enough, when the dogs started baying, she left. She went in a patch of woods and I went with her. Jack Crow, Skinner and Skinner's wife Darlene, kept driving the cows to the pen.

About the time, the old black cow came out on a log road, so did I. I swung my rope one time and chunked. I caught her and stayed behind her until she came to a little opening. I got a rope on her foot, and tied her to a tree. We would take the truck and trailer back and load her.

Jack, Darlene and Skinner had made it almost to the pen when I got there. Skinner was sitting in a hole in the fence where a gate used to be. There was a big high horned cow that just run over his horse and hit the woods. Skinner hollered to me and said, "Catch her!!"

I spun out a loop, and went in the woods behind her. She went through the woods and came out in a long opening. I roped her around the horns but my extra rope was on the old black cow I had tied. I had lost my hat when I came out of the woods and the sun was plenty hot. I got off my horse and tied my reins around my rope, so my horse would face the cow. Then I walked back to get my hat. I looked back, and the cow ran at the horse, and got him wrapped up in the rope, and got him down. The reins was tied to the rope and she had his head pulled around so that he couldn't get up.

I thought someone would be back to help me, so I just set down against a tree in the shade. About an hour later, Jack and Skinner showed up. I said, "Where have you all been? My horse is in a heck of a bind." Jack said they had been eating dinner. Jack asked Skinner if he had a rope in the truck. He said there was one behind the truck seat. Jack got the rope out and got another rope on the cow.

We tied it to the truck and I got my horse untangled. We pulled her over to a tree, and I went and got my truck and trailer and loaded her, and the old black cow. That was the end of he old black cow, because we sold her.

Skinner died not long after that, Darlene has remarried, Jack Crow went back to driving a truck and I kept right on working cattle.

"Cowboying is not a way to make a living. It is a way of life."~Johnny Shroyer

~ CD

WORKING CATTLE

Working cattle is a hard life for the cowboys, horses, and dogs. Through the years I've had several horses hooked by cattle and run a root through one coming up out of an old boggy creek.

I've had lots of dogs killed and broke up by cattle and hogs. Also I've been with people that had dogs killed and horses hooked. Between blow-outs on trucks and trailers, fuel for your truck and keeping it on the road, there's sure not much left for the cowboy.

Ropes, hotshots and other cattle working equipment... Needles and vaccinating gun that you're always breaking... Putting floors in your trailer, repair on your trailers, that you never get paid for. ..
No wonder cowboys are rich.

~ CD

A BRANDON RICHBURG STORY

BIG SHIRLEY AS TOLD BY CD

Shirley had a Brahman heifer that she had raised on a stake rope. The halter ring in the halter wore into after a year or so, and she went into a neighbors pasture. Shirley called Brandon to come catch her. The heifer was gentle, but after Brandon roped her, and drug her into the trailer, she was pretty stirred up. When he got home with her, Shirley had a new halter, and lead rope. She wanted Brandon to put the new halter and lead rope on the heifer for her. By the time he got that done, she was really stirred up.

Shirley got hold of the lead rope, and told Brandon to open the trailer gate. Brandon told her that he didn't think that she could hold the heifer. She told Brandon that she was her baby and she wouldn't be any trouble at all. Brandon opened the gate, and the heifer sailed out in a hard run. They were parked on a gravel oil well pad. It jerked Shirley plum out of her boots. Shirley was a pretty big girl and also pretty big breasted.

It jerked her face down on the gravel flat of her stomach. It yanked all the buttons off her blouse and rolled her bra up under her chin. Shirley's husband was running trying to catch up, but just wasn't fast enough.

When the heifer came by Brandon, he dove on the rope with Shirley. They rolled over three or four times and Brandon's face ended up right between Shirley's breasts. Brandon was pretty embarrassed, but Shirley wasn't.

- CD

Rick Hargrove, Johnny Shroyer, Brandon Richburg, and a few others.

BIG SHIRLEY (BRANDON RICHBURG'S VERSION)

Fifteen years ago there was a man and a woman that bought some land in the Trinity River bottom. They had bought a bottle calf at the sale. A couple of years went by and it grew in to a cow. They kept it staked on an oilwell location. One day it got loose and started living on the side of the road. This went on for a while.

The woman was known as Big Shirley, because she weighed about 350lbs. and was about 5ft tall. They lived in a school bus they had turned into a house.

One day I was driving down the road by their house, and Shirley stopped me. She wanted me to catch her cow, so I told her I would. I unloaded my horse and roped the cow. After loading her in the trailer, I hauled her into an oilwell location. After much trouble, we got the stake rope hooked up to her. By this time, the cow was really mad. Shirley was standing outside the trailer saying, "Settle down, Sugar". So I told Shirley, "You had better move", but she ignored me.

So, as I flung the trailer gate open, Shirley was holding the rope. The cow came out of the trailer at a high rate of speed. As the cow passed Shirley, the rope became half hitched around Shirley's legs. The cow jerked her down and began to drag her in a circle on the gravel location. As Shirley was being drug, she was screaming very loudly. So me and her husband started chasing her.

I cut across and as the cow passed me, I grabbed the rope. The cow was running so hard, the rope was sliding through me hands. Shirley hit me from behind and it knocked me down on top of her. By this time, she was all the way naked. The first thing to hit her was my face, and it was right between her breasts. I guess my mouth was open, cause I noticed her breasts tasted like coconuts.

That cow could not drag me and Shirley together, so she had to stop. But Shirley had wrapped both of her arms around my head and just mashed my face into her breasts.

When the excitement was over, to this day, everytime I smell coconut, I think of Big Shirley. I told my wife about it, and she thought it was sick and hoped I went home and took a bath. **~Brandon Richburg**

ANOTHER BRANDON RICHBURG STORY

Some years back, I had been working all day and had just gotten home, when some people called me needing my help. They had moved all of their cows off a place, but one cow had jumped out of the corral, and they could not get her penned back up. So, they called me to come and get her.

When I got there, they had parents, grandparents, and kids everywhere. It was a big tiger striped cow standing in one corner of the pasture. So, I rode my horse up the side and when she took off, she went right across the middle of the field. I took to her and roped her. About the time I did, I saw a tree go by and I was on one side and she was on the other side and we were flying.

When the rope came tight, I went out over the horse's head with such force I lost both of my boots. One was stuck in the hole behind the saddle horn, and the other was under the cow and she was real mad.

My horse was hung up against the tree. I finally was able to get to him and get back on without the cow getting me. The people brought their trailer out there and I loaded the cow. At that time, I had no boots on, just my socks.

They said they go to the rodeo in Houston every year, and they had never seen someone get off a horse so fast. It seemed to really impress them and they thought that I did it on purpose. They never said anything about me riding my horse in my socks, guess they didn't know it wasn't really supposed to happen that way.

~Brandon Richburg

Brandon Richburg

A BILL PLATT STORY

My friend Bill Platt, Caleb Rhone, and myself was penning cows down at Glen Vaughan's. He told me when we got there, that there was a big steer that jumped out of the lot every time he got him in it. He told me to just rope him and load him in the pasture.

Bill Platt was riding a grey mare that belonged to me. She had a good many cattle roped off her, but not very many big cattle. That was the only big steer left because he had already sold his big calves.

When we got the cattle bunched, I cut the steer out and lined him out down the hill and roped him. Bill came down there where I was and I told him to hold the steer until I got back. He put his rope on him and tied off to the saddle horn. I went to get the truck and trailer, but when I got back I couldn't find him.

There was a marsh up behind a lake with brush and trees all around it. Caleb had heard Bill hollering and he rode down in there to see what was wrong. Bill and the grey mare was on the ground wrapped up in the rope. Bills legs and the mares legs was all tied together. The steer was standing over them making sure they didn't get up.

Bill told Caleb to cut the rope. Caleb asked, "Is that Carroll's rope?" Bill said, "I don't care who's rope it is, cut it!" Caleb cut the rope, and everybody was alright.

P.S. It was Bill's rope that I had tied a horn knot in for him that morning.

~ CD

Bill Platt

A JESS BISHOP STORY

My friend Jess Bishop and myself were working cattle at Frank and Pat Surginer's. We got through a little after dinner and Jess asked me what I was going to do that evening. I told him I didn't have anything planned.

He asked if I would go help him catch a couple of Brahman bulls. He said him and his dad and another man had run them around and around in a hundred acres of woods the day before, but wasn't able to get them.

We went up there, and Jess called his dad, and he met us out there. We were all three horseback, and I sent the dogs into that hundred acres of woods. The dogs found the two bulls together, and they left like a deer. The race was on for about thirty minutes.

We were sitting alongside a fence, when we heard the bulls coming through the brush and the dogs were way behind them. When the bulls got to the fence, they jumped it into another set of woods.

I asked Jess if there was any openings around there that we could get the bulls into. He said there was a gate on the other end of the woods, and if we could get them through the gate, they would be in a big opening. We stayed a long way behind them, and they finally found the open gate by themselves. We waited until they got away from the woods and I built to them.

I ran by the first bull and was going to rope the front bull, and leave the back one for Jess and his dad, but I forgot to tell them the plan. I ran past the back bull, and roped the front one, but the back one was hitting my horse in the back end with his horns. I couldn't slow down, because I was afraid he would job a hole in my horse. I followed the bull until he went through a gate into an old trap. There was a big Oak tree, just after you went through the gate and I ran around the tree with the bull.

Jess and his dad came on down there and Jess got another rope on the bull. I told Jess and Mr. Bishop if they would hold that one, I would go catch the other before he left. The other bull was standing at the other end of the trap, and the dogs were baying on him. When I got nearly to him, he ran at my horse and I jumped behind another big Oak and roped the bull when he came by.

My rope was pretty long, and Mr. Bishop didn't know he would run out of rope before he got to him. His horse made a run for it, but the bull was behind the tree. When the end of the rope came, the horse made an about face, and Mr. Bishop kept going straight ahead. He sure hit the ground hard, but he was okay.

After lots of effort, we finally got the bulls loaded and moved to the river bottom. Jess told his dad if we had this much trouble getting them out of 100 acres, how are we going to get them out of 600 acres? Mr. Bishop said, "Well, now we know the man." **~Jess Bishop**

Jess Bishop & CD Langham

WHERE DID MY HORSE GO?

Some years ago, Carroll Langham and myself worked cattle for people around Houston County. On this particular day, I was to meet Carroll at a restaurant on the north side of Grapeland, TX. for breakfast before we went to a job working cattle.

I arrived at the restaurant with my horse and trailer probably thirty minutes before daylight. I ordered coffee and waited probably ten minutes before Carroll arrived. We ordered breakfast and ate, then sat talking about our day ahead. After awhile we realized that it was good daylight outside, so we paid our bill and went outside.

As we were walking to our trucks, I looked in Carroll's trailer. There wasn't anything in it except his cow dogs. I ask Carroll, "Where's your horse, Carroll?" He looked at me and said, "He's in the trailer." Well I looked again just to make sure I wasn't going blind, and I wasn't because I couldn't see a horse in there.

I told Carroll, "If your horse is in the trailer, show him to me." Well Carroll looked at his trailer, and the look on his face was one that I will remember for the rest of my life.

Now Carroll didn't say a word, but just kept walking toward his truck and trailer. I could tell he was deep in thought. As we walked around behind his trailer there his horse's bridle hung to the back gate on the outside still tied up to the trailer. I said, "Carroll, is that your horse's bridle there? And by the way, where is he?"

Carroll then looked at me and without a smile said, "Well, I caught and saddled my horse before daylight this morning, and tied him here at the back gate of my trailer. I then went and got the dogs from their pens and they ran and got into the trailer.

I got them in the trailer and shut them up behind the front gate, and I guess I just shut the back gate, and went to my truck and drove off. I'll go see if I can find him.

Carroll drove back to his home in Latexo and sure enough his horse was in his pasture with just the saddle on. Carroll loaded him back up and came back to Grapeland.

I asked him where he found his horse and he told me in the pasture. Carroll said, "I guess when I drove off with him tied to the back, he kept up until we came to my cattle guard and he shut it down and pulled his bridle off."

We went on there to work and all day long, Carroll acted as if it was just another day. I'll say it wasn't just another day for me, because every time I looked at Carroll, all day long, I started laughing.

Now this is just one thing that happened to us in our cow working days, that I hold close to my heart, and sometimes when I'm reminiscing about days gone by, I laugh to myself when I remember our cowboy days.

I thank Carroll for a lot of these memories and he remains a close friend and brother to this day. **~Larry Brasher**

Larry Brasher & a young Dr. Clint Warren

ROPING ON THE 7A RANCH

I day worked on the 7A Ranch back in the early 80's. They hired new ranch manager by the name of Bob Thompson. When he came to work there, he brought his horses, dogs, and cattle working equipment with him. He told me that he had a steer release gate and if we had some roping cattle, we could put some roping's on there on the ranch. I told him I would buy some.

Next sale day, I went down to the local livestock auction and bought some native horned cattle that would do to rope. We used one of the bigger pens there on the ranch and set up our head gate. Then, took some Powder River panels and made a little holding pen and alley so we could rope.

We originally planned for mostly the ranch hands, custom hands, and a few locals. But, it turned out to be pretty big. Lots of young people that wanted to rope started coming, and we had a good many ropers, mostly just a community roping. I kept adding more cattle and we had a good many people coming from Leon and Madison County. We had lots of fun and some pretty good wild west wrecks.

A preacher friend of mine, by the name of Jim Bob Gunnels was coming from Leon Co. His little horse didn't want to drag the steer when he got him caught. The horse would just turn and face the steer as soon as he roped him he couldn't turn the horse back around. I told Jim Bob, "Today that horse is going to pull that steer!"

When he got a steer caught, I told one of the other ropers to put his rope on the horse's head so the horse couldn't turn around and face the steer. Well, the horse still wouldn't pull the steer. I got a hotshot and I told Jim Bob to hang on. I touched that horse in the back-end with that hotshot and the rodeo was on. That horse threw Jim Bob three foot above the saddle, and he came down on the saddlehorn right in the seat of his britches. After wallering around on the ground for about five minutes, he said, "Anybody ever have an enema with a saddlehorn?"

Jim Bob was tough, but that horse never did pull a steer. Bob Thompson and myself is too old to rope, but we had lots of fun when we could. I've got lots of stories about that 7A Ranch. I could pretty much fill up a book with them.

Bob has retired and is the Pastor of a church. He's my friend. - C D

A CLEON SALMON STORY

Cleon Salmon had four Brahman cows that was giving him some trouble staying in the pasture. He asked me if I knew anyone that wanted to buy some cows. I told him I may buy them If the price was right. He said lets go look at them, I've got them in the lot and I'm afraid to turn them out, because they will be back on my neighbors and I don't want to lose my good neighbors. We went and looked at them and I bought them.

I asked Cleon if I could use his headgate on his pen to brand them in. He said sure, help yourself. When I went to get them, Gene Allen went over there with me. When we got them branded they were plenty stirred up. We went to load them, they they all went on the trailer except one grey Brahman cow, and all she wanted to do was fight.

I told Gene to get that fence post and whip her on the trailer. He said, "That's your cow, you get that fence post and whip her on the trailer." I got the post and got over in the pen. When my feet hit the ground she came to me. I used the post like a battering ram and hit her between the eyes, but it didn't stop her. That post went plum over the fence! The cow hit me and knocked me under the shed. The post that was holding the barn up, was right there and I'm crawling around it on my hands and knees trying to get away from her.

Finally, I got up on my feet and tried to out run her to the fence. Just as I got to the fence she hit me and threw me over the fence. Gene Allen was laying on the ground laughing . When he finally caught his breath, he said get that post and whip her up on the trailer.

I didn't think it was nearly as funny as he did. - C D

ANOTHER WRECKED SADDLE STORY

Mike Land had a lease pasture west of Lovelady TX. It was over 200 acres. About half of it was open, and about half timber and brush. It was cold in the winter time and the cattle had lice on them and not lots of flesh. He called and wanted me to treat them for lice. He said, "We don't need no help. Me and you can do it if you'll bring me a horse."

The next morning I saddled two horses, loaded my dogs, and went off to work. It was so cold that morning, the water puddles was froze all day long. We got the cattle penned in an old pen made out of pallets and them big spools that light line wire is rolled up on. We started trying to mix that lice off to pour on the cows and the water was so cold it wouldn't gel like it was supposed to. We stirred on it for an hour, and it still wasn't very thick. We could get about three lice cows at a time in the little short chute we had rigged up, and it wasn't easy getting them in there. This one big high horned cow was the only fat cow that he had. I said, "Mike, you need to sell that cow." He said "Why? She is the prettiest cow I've got!" I said "That's because she hadn't had a calf in three years!"

We got the trailer backed up to the chute and went to load her, but she jumped out of the pen. My horse was tied on the outside of the pen, so I jumped on him, spun out a loop, and took to her. She was headed for the woods, which wasn't very far away. I roped her just before we got to the woods and there was a huge Pine tree standing alone before you got to the woods.

We were drifting pretty good when she went behind that Pine tree. I keep going straight, all of a sudden, I'm standing there with my saddle still between my legs, and the horse is still going south. The breast collar and D ring of the saddle was still on the horse. It had ripped the D rings right out through the riggin of the saddle. I took saddle and all and tied her off to the tree. I took my other horse and loaded her.

As I have said before in these stories, "There's not lots of profit in this cowboy work." P.S. Mike didn't tip me for a job well done. ~ C_D

A COWBOY WRECK

James Jackson and myself were working cattle for the Grady Lake Ranch and he wanted to sell some cattle. So, we hauled them in to the East Texas Livestock Cattle Sale. When we got to the sale barn, one of the sale barn owners, Paul Craycraft, asked me if we would run down the road a little ways and pick up some cattle. He said Mr. Barksdale already had them in the lot, but he didn't have a way to haul them.

We told him we would, so we drove over to his place. His pen was just one square pen with a chute in one corner, but no cross fences and no wing or crowd pen to the chute. There was about nine big black Brangus cows in the pen. There was an old aluminum gate laying in the pen and I told James to hold the gate for a wing at the corner of the chute, so I could get on my horse and run them up in the trailer. He agreed to do that, so we put the plan into action.

It looked as though it was going to work, but the front cow hesitated at the back of the trailer. One of the cows from back of the bunch jumped the gate. James landed on his back with the gate on top of him. The whole bunch of cows run right over the top of him and the gate!

Well, "God" looks over fools and he wudin hurt. He told me that's enough of that, you hold the gate. That's when I decided to change plans. We took some old post and sticks and propped the gate up and got the cattle loaded.

That's called "cowboy logic". - CD

CD Langham & Friends

ANOTHER COLORADO STORY

I had a friend by the name of Jack Fleming. We went to Colorado one year elk hunting. Jack Fleming rode one of my horses and he didn't get very far from camp, because the horse figured out he could stop and eat grass anytime he wanted to.

We didn't figure everybody would kill an elk, so we bought only one elk tag and divided the cost between us. All of us bought a mule deer tag, so we were all legal to hunt. We put the elk tag in Jack Fleming's name, because he was first in line where we were getting our hunting license and we had gave him the money to pay. We were camped on the Chama River, just a few miles over the Colorado line, out of New Mexico. We would hunt from there up to the Continental Divide, which was several miles up, and all uphill.

I got in too big a hurry and killed a pretty good bull elk a couple of days early. It was really cold up there, and I skinned him and hung him up on a leaning tree so the bears couldn't eat him. Well, season opened on Saturday. So, Saturday evening, after dark, I rode up leading a pack horse and packed him back down. We had plenty of ice chests and plenty of ice, so I iced him down and put him in some bushes across the Chama River.

The next morning, we were going to carry my elk, and one that Rick Hargrove had killed that morning, over into New Mexico and put them in cold storage. We got the ice chest loaded in the back of a pickup, then they decided to cook breakfast.

Well, while we are cooking breakfast, the Game Warden drove up. He checked the meat and said, this elk was killed Thursday, this thermometer can tell. To make a long story short, it ended up costing me $2300. Since the tag was in Jack Fleming's name, they took his gun, the ice chest, and the elk. Jack had to pay the fine and when I got home, I had to sell eight cows to pay him back.

I loaned him my rifle to hunt with that winter, but he had to go all the way back to Colorado to get his rifle. He was banned from hunting in Colorado for 10 years, plus every state that had elk season.

Not too long after that, a friend of mine by the name of Lacey Lovell was at the East Texas Livestock Auction and he had a friend that was a Colorado Game Warden.

When I walked in to the sale barn cafe, there was Lacey and the Colorado Game Warden in his uniform. Lacey called me over there and introduced me and told me they had more charges and he had come to get me.

It was a dirty joke, but it wasn't funny.

My friend Jack Fleming has finally forgiven me, but it took him lots of years. He is now 93. Thank you brother Jack. ~ CD

Jack Fleming, Gene Allen, CD Langham

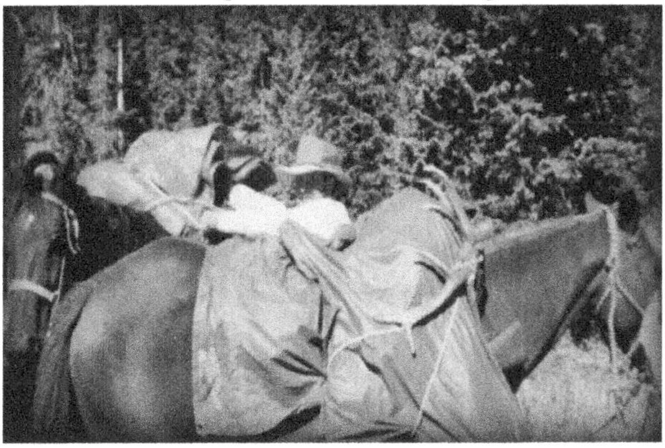

CD Langham packing elk.

EAST TEXAS COWBOYS

If you cowboy in east Texas on the old 7A
You sure better have some dogs that will stay

For the cattle are wild and the brush is thick
And getting the cattle out is a pretty good trick

A good brush horse and a dog that will find
Is the only thing,
That can change them cattle's mind

For out in them thickets, the cattle can sure hide
And you sure better have a good horse to ride

I've had lots of dogs, and some good ones I've had
But a dog that won't handle, can sure make you mad

A dog that will find, and a dog that will stay
Can sure make you have, a much better day

The only thing worse than a dog that won't hunt
Is one that works behind, instead of in front

A boot is a boot, a spur is a spur
But a cow dog's not a cow dog, if it's not a cur

Tell all them hands, out there in the west
That these east Texas cowboys are still the best

- CD

A PENTECOSTAL PREACHER TRAIL RIDE

Two years in a row I put on a trail ride for thirty Pentecostal preachers. I had to furnish the horses, saddles, and wagons for thirty of them, that was a big job. Many of them didn't know how to saddle a horse and most of them hadn't rode a horse in years.

The day before the ride, I would go kill a wild hog and Butch Arnold would carry his BBQ pit where we were going to camp and have supper ready when we got there. Since there was no vehicles involved, they would throw their bedrolls on the wagons and we would spend the night on one of my lease places about 15 miles from where we left from.

The next day was a short ride only about 7 miles back where we left from. We would cook breakfast for them before we left camp. Most of them would be so sore from the day before, that they would walk and lead their horse the seven miles back.

I took every horse in the county to find enough gentle horses for he preachers. We had all of 7A Ranch horses, Doyle Bennett's horses, John Olan Turner's horses, and all of C.D. Langham's horses.

We would have a holy ghost revival that night with all those Pentecostal preachers. It was too much work for me, but it was quite an experience.

~ CD

Prayer before Pentecostal Trail Ride

TRAIL DRIVE - A FAMILY AFFAIR

Dr. Day, a local dentist, had a place fifteen miles from my place. There hadn't been any cattle on the place in a couple of years. I asked him about renting the place for cattle. He told me if I would mow it, I could use it, for nothing. He said the only thing is, there ain't no pens on it and it's for sale, but you can use it until it sells.

I had forty cows at my place and it didn't have much grass left. I loaded them up and hauled them over to that place. It was in the late summer when I carried them over there, and it was knee deep in grass. Well, the place sold about the first of March, and he called me and said I had to move my cattle. It was only fifteen miles from my house, so I decided to drive them home.

So, the 8th day of March, I loaded 5 horses and 3 dogs and headed out. We got over there about 8am. My daughter, who was 14 yrs old, my two sons, 10 and 11 yrs old, my wife, and myself. My daughter was going to drive the truck and trailer ahead of the cattle in case I had to load anything, and my wife and two sons was going to stay on the flank, and I was going to ride drag.

We bayed on the cattle for an hour or so with the dogs so they would be bunched good, and drove them out on the farm road 2022. We hadn't got a mile down the road before it blowed in a wet Norther. It had been a nice day when we left, and none of us had a coat or slicker. We almost froze to death before we got home at 3:30pm. The kids was all crying cause they were so cold, except Athena, who was in the truck with the heater on.

They have lots of memories, but that's not one they will forget. Clint was 10, Gregorios 11, Athena 14, Dorothy 35 and C.D. 37 years old in 1980.

- CD

Family Trail Drive down FM2022

In Honor of Dorothy Langham

A COWBOY'S WIFE

Listen to this story
about a cowboy's wife.
She's got to be a tough one
to live a cowboy's life.

She's got to cook him breakfast
before the morning light.
Then she has to have him supper
when he comes home at night.

Keeps the home fire burning
when the cold wind blows.
Doctor all his cuts and bruises,
wash the cowpoop off his clothes.

When a cowboy's day is over,
his horse and dogs are fed.
Then she's still up cleaning,
when he's done gone to bed.

Feeding all the livestock
that he has in the pen.
Raising all the orphans,
that he comes dragging in.

Listen to his old stories
that he's already told,
as he tells about the weather
and how it's been so cold.

So when he gets home at night
and a cowboy's work is done
Think about that cowboy's wife,
her work has just begun.

For all the work she did here
I know she will be blessed.
And now that she's in heaven,
she can get some rest.

LIVING IN THE PAST

I wish that I had been here
When the land was wild
No fences to restrain me
And you could ride for a while

The horses had more bottom
And the cattle all had horns
These days that I dream of
Is when I wish that I was born

A fast horse and a long rope
Would be the tools of my trade
The mavericks that I branded
Would be the money that I made

They say that I was born too late
For the life that I've lived
But I'd give anything to have been there
If I had anything to give

Now the yard lights are always shining instead of the stars
And you can't even ride your horse, because of all the cars
My horse's feet need trimming, their hooves are grown out long
My saddles all need oiling, but I don't know what is wrong

All I do is sit around and think of the days gone by
The next stop is the old folks home
Just waiting there to die.

This book that you've been reading, is my first and my last
These stories that I've written are the stories of my past
When I'm sitting in that old folks home, with a feeble mind
Read me all these stories, because I'll probably be too blind

~ CD

OLD HORSES

My horses are all getting old, and they're not too sound
Weather knots in their mane, their tails near drag the ground

If they had to make a day, I doubt if they could last
If they had to run at all, they couldn't run too fast

I've had lots of horses, and some of the best
Now they're turned out to pasture, all they do is rest

They made lots of days, and some were long and hard
But they sure do look bored, now that they're retired

I wonder if they do like me, and dream of yesterday
Or maybe they're just standing there, and waiting on some hay

~ CD

Richard Orr (Nephew of CD Langham)

IN MEMORY OF RICHARD ORR

Richard was my nephew. When he was growing up, he wanted to do everything that I did. He loved being horseback, and we spent lots of time in the company woods that we ran cattle in. He went to Colorado twenty-eight years ago, where he worked as an outfitter packing tourists and hunters back in the mountains.

He lived on Lake Vallecito most of the time he was there, and up until he passed away in November 2015, at 56 years old. We had a memorial for him on June 4th of 2016. They shot his ashes out of a cannon over the lake. Richard had many friends that gave him a big going away party which we attended in Colorado. I wrote this poem to read at his memorial.

RICHARD

Richard was a wild one,
He loved being free,
And up there in the mountains,
is where he'd rather be.

In this old world,
there's not much place to hide,
But you sure can feel hid,
up on that Great Divide.

Just one great big adventure,
that's how he lived his life,
He sure didn't have much time,
for a home and a wife.

When he was just a youngun,
and I was in my prime,
Riding in them big woods,
is where we spent our time.

He left this old world early,
but no reason to cry,
He done lots of living,
before he said goodbye.

When he took his last trip,
to that far away bye and bye,
I hope "God" gave him a home in the mountains,
instead of a mansion in the sky.

"Love," Uncle Buddy

TRICK ROPING

I was invited to do some trick roping throughout my life at the Latexo Elementary School a couple of times, and also for some childrens' bible schools. These are some of the pictures they took and sent me copies of. The kids and me had fun. - C D

THE END

My old hoss is turned out to pasture
He just can't understand
Instead of handling cattle
I handle these pots and pans

Instead of ketchin wild ones
I'm cooking these here beans
But I'm still ketchin wild ones
Only I ketch-um in my dreams

I don't do no more cowboyin
That's for younger men
But I'm still making stories
Only I make-um with this pen

I don't know what's in the future
Or what I have in store
But I ain't gonna rope the wild ones
Or ride the rough ones anymore

~ CD

TIME WAITS FOR NO ONE

Our days on earth are numbered, life's sun is sinking low

Time waits for no one, and soon it's time to go

So while were here and waiting, make sure our work is done

Cause time waits for no one, like the setting of the sun

So let's don't fail to witness, this story must be told

Our spirits must stay warm, or our hearts will grow cold

Let us show the love "God" gave us, let us use our gifts each day

Cause time waits for no one, let us use this time to pray

Don't worry about our bodies, just worry abour our souls

Cause time waits for no one, and soon we will be old

Our lives should not be wasted, "God" put us here to use

We have to win this battle, with "God" we can not lose

So, daily work for "Jesus", let our lights shine bright.

Cause time waits for no one, and soon it will be night

Put on all our armour, when we fight against sin

Causes time waits for no one, we have one chance to win

Time waits for no one, we have one chance to win

- CD

These are the cowboys that I've worked with through the years, not including my wife and kids.

David Harkins
Walter Sherrouse
Billy Bob Christian
Darrell Gilmore
Bill Merchant
Mark Avery
Todd Merriwether
Cliff Pyburn
Rick Hargrove
Butch Arnold
Jimmy McLean

L.D. Patton
Billy Crowson
Ronnie Goolsby
Terry Huddleston
Jess Bishop
Larry Brasher
James Jackson
Justin Baker
Juan Hernandez
James Ashley
Caleb Rhone

And probably some more I left out like:

Alfredo Fajardo
James Lloyd Murray
Bob Thompson
Jack Crow
Perry Jones
Dickey Manvel
Matt Benton
Earl Dean Benton
Brandon Richburg
Carl Murphy

Bill Platt
Sammy Lane
Henry Beaver
Billy Davenport
Jason Forum
Joe O'Berry
Jaret McDaniels
June Bug Reese
Jimmy Smith
Ronnie Murphy

Made in the USA
Coppell, TX
07 June 2024

33218958R00066